D1175226

Medical College
of Pennsylvania and
Hahnemann University

gratefully acknowledges
a contribution in
support of library resources
by

DR. PHILIP ROSENBERG

**MEDICAL
COLLEGE OF
PENNSYLVANIA AND
HAHNEMANN
UNIVERSITY**

FORMALDEHYDE
TOXICITY

CHEMICAL INDUSTRY INSTITUTE OF TOXICOLOGY SERIES

Gibson: Formaldehyde Toxicity

Golberg: Structure-Activity Correlation as a Predictive Tool in Toxicology

Gralla: Scientific Considerations in Monitoring and Evaluating Toxicological Research

Forthcoming

Hamm: Complications of Viral and Mycoplasmal Infections in Rodents to Toxicology Research and Testing

Popp: Current Perspectives in Mouse Liver Neoplasia

Rickert: Toxicity of Nitroaromatic Compounds

FORMALDEHYDE TOXICITY

Edited by

James E. Gibson

Chemical Industry Institute of Toxicology

⬤HEMISPHERE PUBLISHING CORPORATION

Washington   New York   London

DISTRIBUTION OUTSIDE THE UNITED STATES
McGRAW-HILL INTERNATIONAL BOOK COMPANY

Auckland   Bogotá   Guatemala   Hamburg   Johannesburg   Lisbon
London   Madrid   Mexico   Montreal   New Delhi   Panama   Paris
San Juan   São Paulo   Singapore   Sydney   Tokyo   Toronto

This book was set in Press Roman by Hemisphere Publishing Corporation. The editors were Christine R. Flint and Sandra J. King; the production supervisor was Miriam Gonzalez; and the typesetter was Wayne Hutchins.
Braun-Brumfield, Inc, was printer and binder.

FORMALDEHYDE TOXICITY

Copyright © 1983 by Hemisphere Publishing Corporation. All rights reserved. Printed in the United States of America. Except as permitted under the United States Copyright Act of 1976, no part of this publication may be reproduced or distributed in any form or by any means, or stored in a data base or retrieval system, without the prior written permission of the publisher.

1 2 3 4 5 6 7 8 9 0   B R B R   8 9 8 7 6 5 4 3

Library of Congress Cataloging in Publication Data
Main entry under title:

Formaldehyde toxicity.

  (Chemical Industry Institute of Toxicology series)
  Includes bibliographical references and index.
  1. Formaldehyde—Toxicology.   I. Gibson, James E.,
date. II. Series. [DNLM: 1. Formaldehyde—
Toxicity—Congresses. QV 225 F723 1980]
RA1242.F6F67   615.9'5136   82-6189
ISBN 0-89116-275-5   AACR2
ISSN 0278-6265

Contents

PART 2: TOXICOLOGY

PART 3: HUMAN STUDIES

Contributors

RONALD AIRMAN, M.D.
New Jersey State Deparement
   of Health
Trenton, New Jersey

IB ANDERSEN, M.D.
Institute of Occupational Health
Hellerup, Denmark

JOHN ASHBY, Ph.D.
Imperial Chemical Industries, Ltd.
Alderley Park, Cheshire, United
   Kingdom

CRAIG S. BARROW, Ph.D.
Chemical Industry Institute
   of Toxicology
Research Triangle Park, North Carolina

MARIO C. BATTIGELLI, M.D.,
   M.P.H.
University of North Carolina
Chapel Hill, North Carolina

PATRICIA B. BLUNDEN, B.S.P.H.
Chemical Industry Institute
   of Toxicology
Research Triangle Park, North Carolina

HENRY F. BOLTE, D.V.M., Ph.D.
Bio/dynamics, Inc.
East Millstone, New Jersey

CRAIG J. BOREIKO, Ph.D.
Chemical Industry Institute
   of Toxicology
Research Triangle Park, North Carolina

DAVID J. BRUSICK, Ph.D.
Litton Bionetics, Inc.
Kensington, Maryland

JANE C. F. CHANG, Ph.D.
Chemical Industry Institute
   of Toxicology
Research Triangle Park, North Carolina

FRED CHASALOW, Ph.D.
Washington University
St. Louis, Missouri

THEODORE Y. CHIN, Ph.D.
Stauffer Chemical Company
Farmington, Connecticut

JOHN J. CLARY, Ph.D.
Celanese Corporation
New York, New York

R. DANIEL DAL CORSO, M.S.P.H.
Chemical Industry Institute
   of Toxicology
Research Triangle Park, North Carolina

DAVID J. DONOFRIO, D.V.M.
Battelle Columbus Laboratories
Columbus, Ohio

DONALD B. FELDMAN, D.V.M.
Research Triangle Institute
Research Triangle Park, North Carolina

JOSEPH F. FRAUMENI, JR., M.D.
National Cancer Institute
Bethesda, Maryland

JOHN R. FROINES, Ph.D.
School of Public Health
University of California, Los Angeles

JOHN F. GAMBLE, Ph.D.
Appalachian Laboratory
   for Occupational Safety and Health
Morgantown, West Virginia

DAVID W. GAYLOR, Ph.D.
National Center for Toxicological
   Research
Jefferson, Arkansas

JAMES E. GIBSON, Ph.D.
Chemical Industry Institute
   of Toxicology
Research Triangle Park, North Carolina

LEON GOLBERG, M.B., D.Sc., D.Phil.
Duke University Medical Center
Durham, North Carolina

RICHARD A. GRIESEMER, D.V.M.,
   Ph.D.
Oak Ridge National Laboratory
Oak Ridge, Tennessee

ELIZABETH A. GROSS, B.A.
Chemical Industry Institute
   of Toxicology
Research Triangle Park, North Carolina

HENRY D'A. HECK, Ph.D.
Chemical Industry Institute
   of Toxicology
Research Triangle Park, North Carolina

A. ROBERT JEFFCOAT, Ph.D.
Research Triangle Institute
Research Triangle Park, North Carolina

WILLIAM D. KERNS, D.V.M., M.S.
Smith, Kline & French
Philadelphia, Pennsylvania

PAUL LEFEVRE, Ph.D.
Imperial Chemical Industries Limited
Alderley Park, Cheshire, United
    Kingdom

RICHARD J. LEVINE, M.D.
Chemical Industry Institute
    of Toxicology
Research Triangle Park, North Carolina

HOWARD I. MAIBACH, M.D.
University of California Medical Center
San Francisco, California

HARRY MARR, B.S.
Research Triangle Institute
Research Triangle Park, North Carolina

GARY M. MARSH, Sc.D.
University of Pittsburgh
Pittsburgh, Pennsylvania

JOSEPH MARTIN, EMT (EMSA)
Chemical Industry Institute
    of Toxicology
Research Triangle Park, North Carolina

LARS MØLHAVE, M.Sc.
University of Aarhus
Aarhus, Denmark

GORDON W. NEWELL, Ph.D.
Electric Power Research Institute
Palo Alto, California

KENNETH L. PAVKOV, Ph.D.
Battelle Columbus Laboratories
Columbus, Ohio

JAMES A. POPP, D.V.M., Ph.D.
Chemical Industry Institute
    of Toxicology
Research Triangle Park, North Carolina

DANIEL L. RAGAN, B.S.
University of North Carolina
Chapel Hill, North Carolina

WILLIAM E. RINEHART, Sc.D.
Allied Chemical Corporation
Morristown, New Jersey

GEORGE M. RUSCH, Ph.D.
Allied Chemical Corporation
Morristown, New Jersey

MERCEDES C. SCHMITZ, B.S., M.S.
Chemical Industry Institute
    of Toxicology
Research Triangle Park, North Carolina

ROBERT E. STAPLES, Ph.D.
E. I. du Pont de Nemours & Co., Inc.
Newark, Delaware

WILLIAM H. STEINHAGEN
Chemical Industry Institute
    of Toxicology
Research Triangle Park, North Carolina

JAMES A. SWENBERG, D.V.M., Ph.D.
Chemical Industry Institute
    of Toxicology
Research Triangle Park, North Carolina

MICHAEL J. THUN, M.D.
National Institute of Occupational
    Safety & Health
Cincinnati, Ohio

JUDY WALRATH, Ph.D.
National Cancer Institute
Bethesda, Maryland

OTTO WONG, Sc.D.
Environmental Health Associates, Inc.
Berkeley, California

MARY ANN WOODBURY, M.C.H.
University of Cincinnati
Cincinnati, Ohio

CARL ZENZ, M.D.
University of Wisconsin
Madison, Wisconsin

Foreword

This volume discusses the toxicology of formaldehyde. There are many reasons why formaldehyde plays a key role in industrialized societies, just as it does in the metabolic machinery of all, or virtually all, living creatures. There is a fundamental reason for this, since we are dealing here with the most reactive of the one-carbon compounds, a molecule that polymerizes with itself spontaneously as well as undergoing a host of other condensation, oxidation, reduction, and alternative reactions. In fact, there is something so primeval about formaldehyde that it comes as no surprise to find it existing in "vast abundance" in interstellar space (1), along with NH_3, HCN, HCNO, and similar as well as derived reactive species. Formaldehyde is associated also with inorganic dust clouds in space. Equally, it stands to reason that a key part would be played by formaldehyde in the abiotic synthesis of biochemically important compounds, and thus in the origin of life on this planet (2-11).

The ubiquity of formaldehyde in the natural and manmade environment (12-14) is matched only by its importance as an endogenous metabolite. Like Salt Lake City situated at the crossroads of the West, formaldehyde sits astride the metabolic pathway of one of the body's key units of currency, the methyl group. The ramifications of the aldehyde's reactivity have not yet been fully explored, so that suggestions of harmful effects stemming from excessive amounts of endogenously generated formaldehyde remain hypothetical possibilities (15) that have not penetrated the consciousness of the physician.

From the A.C.S. Monograph on formaldehyde (4) we learn that it was in 1859 that A. M. Butlerov first described various forms of formaldehyde and its reactions, including the formation of hexamethylenetetramine. Nine years later, A. W. Hoffmann prepared formaldehyde from methanol and characterized it definitively as the first member of the homologous series of aldehydes. Formaldehyde has been the subject of a number of monographs, principally devoted to its chemistry (Table 1).

The toxicology of formaldehyde, particularly with reference to the long-term effects of vapor inhalation, has received scant attention until just recently. In fact, it was recognition by the chemical industry in this country of the urgent need to remedy such deficiencies that led to the establishment of CIIT. Like any new boy, CIIT started at the bottom of the ladder, working with the first members of several homologous series: alkenes (ethylene), alkyl halides (chloromethane), aldehydes (formaldehyde), secondary aliphatic amines (dimethylamine), aromatic hydrocarbons (benzene, toluene), and aromatic amines (aniline). Even that hoary old-timer chlorine, discovered by Scheele in 1774, is only now being studied with respect to its long-term effects. On our priority list are similar "firsts"—both organic and inorganic—that we hope to address in the near future. Let me stress that the decisions on priority compounds, and on the order and manner of testing, were made by CIIT quite independently of its member companies, just as CIIT plans the remainder of its research program.

This book serves to emphasize the importance of CIIT's mission. The structure of this volume illustrates CIIT's approach, which encompasses state-of-the-art testing, studies to achieve understanding of mechanisms of toxic action, evidence derived from human exposure, and the final application of all these data to the assessment of risk. We owe a debt of gratitude to the Committee responsible for organizing the Conference from which this book derived: Dr. J. E.

Table 1   Monographs on Formaldehyde[*]

Year of publication	Author	Country
1908	J. E. Orlov	Russia
1909 (Transl.)	D. Kietabl	Germany
Early 1900s	L. Vanino and E. Seitter	Germany
1927 (Revision)	A. Menzel	Germany
1944 (1st ed.)	J. F. Walker	U.S.A.
1953 (2d ed.)	J. F. Walker	U.S.A.
1964 (3d ed.)	J. F. Walker	U.S.A.
1975 (3d ed. reprint)	J. F. Walker	U.S.A.

[*]Summarized from Walker, JF. *Formaldehyde*. Huntington, New York: Robert E. Krieger (1975).

Gibson (Chairman), Drs. R. J. Levine and J. A. Swenberg of CIIT, Dr. J. J. Clary (Celanese), and Dr. C. F. Reinhardt (Du Pont).

REFERENCES

1 Snyder, LE, D Buhl, B Zuckerman, and P Palmer. Discovery of formaldehyde in interstellar space. *Phys. Rev. Letters* 22:679 (1969).

2 Pavlovskaya, TE. Possible ways of identifying the abiogenesis of biochemically important compounds. In *The Origin of Life and Evolutionary Biochemistry*, edited by K Dose, SW Fox, GA Deborin, and TE Pavlovskaya, pp. 387–395. New York: Plenum Press (1974).

3 Calvin, M. View from the past towards the present. In *Chemical Evolution*, pp. 124–129. New York: Oxford University Press (1969).

4 Pinto, JP, GR Gladstone, and YL Yung. Photochemical production of formaldehyde in earth's primitive atmosphere. *Science* 210:183–185 (1980).

5 Decker, P, and A Speidel. Open systems which can mutate between several steady states ("bioids") and a possible prebiological role of the autocatalytic condensation of formaldehyde. *Z. Naturforsch. [B.]* 27:257–263 (1972).

6 Becker, RS, T Bercovici, and K Hong. New reactions of paraformaldehyde and formaldehyde with inorganic compounds. *J. Mol. Evol.* 4:173–178 (1974).

7 Caballol, R, R Carbó, R Gallifa, JA Hernández, M Martin, and JM Riera. Theoretical interstellar and prebiotic organic chemistry: A tentative methodology. *Origins of Life* 7:163–173 (1976).

8 Decker, P. Did evolution begin with formaldehyde. *Umschau* 73:733–734 (1973).

9 Kamaluddin, X, H Yanagawa, and F Egami. Formation of molecules of biological interest from formaldehyde and hydroxylamine in a modified sea medium. *J. Biochem. (Tokyo)* 85:1503–1507 (1979).

10 Zdzislaw, I. Cosmic seeds of life. *Postepy Astronaut.* 12:25–35 (1979).

11 Yanagawa, H, Y Kobayashi, and F Egami. Genesis of amino acids in the primeval sea. Formation of amino acids from sugars and ammonia in a modified sea medium. *J. Biochem. (Tokyo)* 87:359–362 (1980).

12 Dhar, NR, and A Ram. Formaldehyde in the upper atmosphere. *Nature* 132:819–820 (1933).

13 Committee on Medical and Biologic Effects of Environmental Pollutants. Sources of atmospheric hydrocarbon. In: *Vapor-phase Organic Pollutants*, Washington, D.C.: National Academy of Sciences (1976).

14 Committee for the Working Conference on Principles of Protocols for Evaluating Chemicals in the Environment. Atmosphere. In *Principles for Evaluating Chemicals in the Environment*, pp. 381–385. Washington, D.C.: National Academy of Sciences (1975).

15 Barker, SA, GF Carl, and JA Monti. Hyperformaldehydism: A unifying hypothesis for the major biochemical theories of schizophrenia. *Med. Hypotheses* 6:671–686 (1980).

Leon Golberg

Preface

The purpose and objective of the Conference from which this book derived was to bring together current knowledge of the known and potential toxicity of formaldehyde, and to make suggestions for future studies in order to make possible scientific risk assessments for humans under actual use conditions. The Conference therefore identified "knowledge gaps" concerning formaldehyde's toxicity.

Formaldehyde is a normal metabolic product, produced in most life forms as a vital precursor in the synthesis of other biochemicals essential to life.

It is also an important commodity chemical, with a worldwide volume of over 10 billion pounds per year. The manufacture of urea-formaldehyde resins, phenolic resins, polyacetal resins, pentaerythritol, hexamethylenetetramine, and melamine-formaldehyde resins accounts for the majority of uses of formaldehyde. The commonplace use of these materials by modern society indicates their importance in our daily lives.

Despite these features of formaldehyde's occurrence and use, there is little information available on the chronic toxicity and potential carcinogenicity of this valuable chemical.

To counter this deficiency, CIIT initiated in June of 1978 a long-term toxicity and carcinogenicity study in Fischer-344 rats and B6C3F1 mice using inhalation exposure. After 18 months of exposure, 6 h/day, 5 days/week, a high incidence of squamous cell carcinomas had been diagnosed in the nasal turbinates of rats exposed to 15 ppm formaldehyde. These tumors had not occurred at 6 or 2 ppm, nor in mice exposed to 2, 6, or 15 ppm formaldehyde vapor.

In early June 1980, formaldehyde inhalation exposures to rats and mice were discontinued, according to CIIT protocol, and the majority of the animals were examined for changes in clinical chemistries, urinalyses, hematology, gross pathology, and histopathology. Some animals were retained, without formaldehyde exposure, for similar studies at 27 and 30 months after study initiation.

To provide for the rapid dissemination of the results of these CIIT findings with formaldehyde, CIIT decided to make formaldehyde toxicity the subject of its Third CIIT Conference on Toxicology. All available data from the CIIT long-term toxicity and carcinogenicity study of formaldehyde in rats and mice were presented. In addition, CIIT presented results of its ancillary research bearing on sensory irritation, inhalation disposition, cell transformation, and mechanisms of toxicity. Completed and in progress epidemiology findings were also presented.

Testing and research results from other laboratories were included to provide a complete and well-balanced consideration of the subject. Experts, representing several scientific disciplines, were invited to participate in the formal program. In many cases the manuscripts contained herein have been updated to include the results presented in the final report of the CIIT study.

James E. Gibson

Part One

Overview

Chapter 1

Overview of Formaldehyde

Gordon W. Newell

It was only a few years ago that the innovative urea-formaldehyde foam process was introduced as a new system for insulating residences. Shortly thereafter, however, complaints about formaldehyde odors in certain residences treated by this process prompted a flurry of investigations into the problem. Subsequent concerns about irritation and headaches in some of the exposed populations were compounded by a report from the Chemical Industry Institute of Toxicology: in a CIIT-sponsored inhalation study, rats developed squamous cell carcinomas in the nasal cavity when exposed for long periods to 15 ppm concentrations of formaldehyde.

In this presentation I will touch briefly on the chemistry, occurrence, biochemistry, environmental/occupational exposure, and health effects of formaldehyde. Much of the information has been taken from two National Academy of Sciences publications: "Formaldehyde—An Assessment of Its Health Effects," published in March (1980) by the Committee on Toxicology, and a report of the Committee on Aldehydes, "Formaldehyde and Other Aldehydes," published in December (1981).

PROPERTIES OF FORMALDEHYDE

Monomeric formaldehyde is a colorless gas that condenses to form a liquid of high vapor pressure that boils at $-19°C$; it forms a crystalline solid at $-118°C$. It has a pungent odor that is highly irritating to the exposed membranes of the eyes, nose, and upper respiratory tract. In the pure, dry, liquid form at low temperatures (-90 to $-117°C$) it polymerizes slowly; its stability depends on its purity, and it must be held at low temperature to avoid polymerization.

Trioxane is the cyclic trimer of formaldehyde; it has the empirical formula of $C_3H_6O_3$, with three formaldehyde units per molecule. In pure form, it is a colorless, crystalline solid that melts at $61-62°C$ and boils at $115°C$. It has a chloroformlike odor but is not irritating. It is soluble in water but is combustible and burns readily when ignited.

Paraformaldehyde contains eight structural units of formaldehyde. It is a colorless solid in a granular form with an odor characteristic of monomeric formaldehyde. Paraformaldehyde melts over a wide temperature range, which depends on the degree of polymerization. At room temperatures, it gradually vaporizes largely as a monomeric formaldehyde with some water formation; the rate is increased by heating. Thus, it is commonly used as a source of formaldehyde for disinfecting large areas.

Formalin is the principal form in which formaldehyde is marketed; it is an aqueous solution that ranges in concentration from 37 to 50 percent by weight. It is a clear solution with a strong, pungent odor of formaldehyde. The solution is slightly acidic.

Formaldehyde vapor is relatively stable with respect to thermal decomposition; at temperatures above $400°C$, it decomposes to form carbon monoxide, hydrogen, and methanol. In the aqueous phase, formaldehyde is oxidized readily by even mild oxidizing agents; this property has been exploited in the development of several wet, chemical, analytical methods for formaldehyde.

When oxidized under controlled conditions, in the gaseous or dissolved state, formaldehyde may be converted in part to formic acid or, under more highly oxidative conditions, to carbon monoxide and water. The photooxidation of formaldehyde in the gas phase leads to carbon monoxide, hydrogen, hydrogen peroxide, formic acid, and some other meta-stable products.

The formation of resinous products on reaction with other chemicals is one of the most useful characteristics of formaldehyde and is the reason for its immense importance in the synthetic resin industry. Under suitable conditions, the molecules of many compounds are linked together by methylene groups when subjected to the action of formaldehyde. Phenol- and urea-formaldehyde resins are polymethylene compounds of this type. Two distinct mechanisms are probably involved in resin-forming reactions: the polycondensation of simple methyl derivatives and the polymerization of double-bonded methylene compounds.

Thermoplastic resins are the result of simple linear condensations, whereas the production of thermosetting resins involves the formation of methylene cross-linkage between linear chains. Both types may be produced from the same raw materials by variations in the relative amounts of formaldehyde used, the condition of the catalysts, and the temperature. However, with compounds whose molecules present only two reactive hydrogen atoms, only thermoplastic resins can be obtained.

SOURCES AND OCCURRENCES OF FORMALDEHYDE

Formaldehyde emissions from industrial processes are generally confined to the immediate vicinity of the plant. Primary sources of potential public exposure include cigarette smoke, automotive exhaust, photochemical smog, incinerators, and degassing of urea-formaldehyde resinous products (industrial processes).

Incomplete combustion of hydrocarbons accounts for much of the formaldehyde present in the atmosphere. Automotive exhausts have been reported to contain formaldehyde at 29-43 ppm (1). Mobile sources (automobiles, diesel engines, and aircraft engines) emit approximately 666 million pounds of formaldehyde annually. Local concentrations may vary with traffic patterns and vehicular density.

Municipal incinerators emit about 0.6-0.9 pound of formaldehyde per ton of refuse, or 13.1 million pounds of formaldehyde annually (2).

Photochemical smog can be an important source of formaldehyde. Stupfel (3) reported that outdoor air in Los Angeles contained formaldehyde at 0.05-0.12 ppm over the course of 26 days of measurements. A heavy smoker can be exposed to a considerable amount of formaldehyde. Cigarette smoke contains as much as 40 ppm of formaldehyde by volume (4, 5). With 95 percent retention from ten 40-ml puffs on each of 20 cigarettes, a smoker could receive a total daily burden of approximately 0.4 mg of formaldehyde.

Other potential sources of formaldehyde in the home include combustion and gas stoves, heaters, and breakdown of cooking oils. Hallowell et al. (6) measured total aliphatic aldehydes in indoor air. After human occupation of an experimental house, total aldehydes increased threefold to 0.116 ppm. The ventilation rate was only 0.02 air exchanges each hour.

A recent study of 15 occupied residential units revealed formaldehyde concentrations of less than 0.12 ppm in 11 of the units (7). Concentrations as high as 0.38 and 0.31 ppm were found in two of the units, which were mobile homes containing particleboard. Formaldehyde concentrations of 0.03-2.5 ppm were measured in 74 mobile homes whose occupants complained of odor and irritation thought to be associated with the use of particleboard (8). Repeat measurements on two homes indicated half-lives (time for concentrations to decrease by 50 percent) of 45 and 110 days. A Scandinavian study using field tests and mathematical models indicated a half-life of 2 years (9). The rate of formaldehyde disappearance therefore may depend heavily on air temperature, ventilation rates, surface areas of the various products, type of material, and volume of the residence.

METABOLISM AND BIOCHEMISTRY

Formaldehyde is a normal metabolite in mammalian systems and, in small quantities, is rapidly metabolized (10). The major route of biotransformation appears to be oxidation to formic acid, followed by further oxidation to carbon dioxide and water (11, 12). In humans, the formation of formate from formaldehyde appears to involve an initial reaction with glutathione to form a hemiacetal (13). The enzyme formaldehyde dehydrogenase then oxidizes the hemiacetal to formic acid with nicotinamide adenine dinucleotide (NAD) as a hydrogen acceptor (14). In humans, formaldehyde dehydrogenase is a multifunctional complex of enzymes that converts methanol to formic acid without releasing formaldehyde as an intermediate, inasmuch as it is difficult to isolate the enzyme alone (13, 14).

Administration of radiolabeled formaldehyde to rats by the oral or intraperitoneal route resulted in 40 and 82 percent, respectively, of the label being found in respiratory carbon dioxide (12, 15).

Numerous enzymes capable of catalyzing the reaction of formaldehyde to formic acid have been identified in liver preparations and erythrocytes (16, 17). The adverse effects of formaldehyde may be related to this high reactivity with the amines and formation of methylol adducts with nucleic acids, histones, proteins, and amino acids. The methylol adducts can react further to form methylene linkages among these reactants (18). It appears that before formaldehyde reacts with amino groups in RNA, the hydrogen bonds forming the coiled RNA are broken down. Formaldehyde reacts with DNA less frequently than with RNA because the hydrogen bonds holding DNA in its double helix are more stable (19). Reaction of formaldehyde with DNA has been observed, by spectrophotometry and electron microscopy, to result in irreversible denaturation. If permanent cross-links are formed between DNA reactive sites and formaldehyde, these links could interfere with replications of DNA and may result in a mutation.

OCCUPATIONAL EXPOSURE

Workers in plants producing plywood or particleboard often use either urea-formaldehyde, phenol-formaldehyde, or melamine-formaldehyde resins. The formaldehyde levels in several such plants have been reported to be as high as 10 ppm (20-22). The formaldehyde levels in the air inside these plants obviously depend on the ventilating system. Other key variables include the amount of free formaldehyde in the resin, the moisture content of the wood, humidity of the air inside the plant, and the processing temperatures. With current emission control technology, significantly lower formaldehyde levels are or can be obtained.

The 1976 National Institute for Occupational Health and Safety (NIOSH) "Criteria for a Recommended Standard . . . Occupational Exposure to Formaldehyde" identified over 60 occupations wherein workers could potentially be exposed to formaldehyde. These ranged from anatomists and agriculture workers to varnish workers and wood preservers (23).

HEALTH EFFECTS

The toxicological effects of formaldehyde on experimental animals will be considered in detail elsewhere in this volume and thus will not be considered here.

Only a few controlled exposure studies with humans have provided useful dose-response data on the irritant effects of airborne formaldehyde. One such study involved 16 healthy young Danish subjects exposed to formaldehyde at concentrations of 0.25, 0.42, 0.83, or 1.6 ppm for 5 h/day for 4 days (24). No significant changes were observed in pulmonary function, nor was there any difference in performance of mathematical tests between the control period and exposure to formaldehyde. Mucus flow rate was decreased at all concentrations except for the 0.83 ppm group. Comments by the exposed groups included complaints of conjunctival irritation and dryness of the nose and throat.

In a second study, 33 German subjects were exposed to formaldehyde at concentrations in a range of 0.03–3.2 ppm for a total of 35 min. Forty-eight others were exposed at 0.03–4.0 ppm for 1.5 min (25). Responses measured included eye, nose, and throat irritation, odor, "desire to leave the room," and increases in the eye blinking rate. An approximately linear relationship was found for the average responses over the range of concentrations. In another study, eye irritation responses to exposures to formaldehyde for 5 min, at concentrations ranging from 0.01 to 1.0 ppm, were investigated in 12 subjects (26). Between 0.3 and 1 ppm there was a linear increase in the average reported eye irritation response. At concentrations below 0.3 ppm a linear relationship was not found.

Acute ingestion of Formalin by humans has resulted in loss of consciousness, vascular collapse, pneumonia, hemorrhagic nephritis, and abortion. Formaldehyde has occasionally injured the larynx and trachea, but damage to the gastrointestinal tract occurred primarily in the stomach and lower esophagus.

Contact of the skin with formaldehyde may cause primary irritation or allergic dermatitis (27, 28). There have been reports of sensitivity to Formalin in nurses who handled thermometers that had been immersed in a 10% solution of formaldehyde (29). A similar response occurred in a hemodialysis unit where a 2% Formalin solution was used to sterilize open tanks (30).

Formaldehyde has been shown to be a potent experimental allergin in humans. In a group of male subjects given repeated occlusive applications of 5% or 10% aqueous formaldehyde for $3\frac{1}{2}$ weeks and then challenged with 1% applications 2 weeks later, some 8 percent of the subjects developed skin sensitization (31).

Experiments suggest, however, that most sensitized subjects can tolerate exposure to aqueous formaldehyde at 30 ppm applied to the axilla (32, 33). Formaldehyde acts as a mucous membrane irritant to cause conjunctivitis and lacrimation. Eye irritation is a common complaint and has been reported at airborne concentrations of 0.3-0.9 ppm in industrial workers (34). Formaldehyde has been reported to cause irritation and dryness of the nose and throat and olfactory fatigue. Upper airway irritation attributed to formaldehyde at 1-11 ppm occurred in employees handling nylon fabric coated with urea-formaldehyde resin (35). Lower airway irritation may be evidenced by cough, chest tightness, and wheezing. One man developed dyspnea and asthma after acute inhalation of Formalin vapor (36). Clinical examination revealed pulmonary edema with a 40 percent decrease in vital capacity (Table 1).

SUMMARY AND RECOMMENDATIONS

The health effects associated with exposure to formaldehyde cover a wide range of signs and symptoms. Most are related to the irritating properties of formaldehyde involving the eyes, nose, and throat. The severity of response is related to exposure concentration and can vary from person to person. Responses may be categorized as follows:

1 Those that produce discomfort (irritation)
2 Those that result in more significant effects, such as increased airway resistance and severe tissue damage in the respiratory tract
3 Those that result in sensitization

Table 1   Health Effects of Formaldehyde at Various Concentrations

Reported effects	Formaldehyde concentration (ppm)
None reported	0-0.05
Neurophysiological effects	0.05-1.05
Odor threshold	0.05-1.0
Eye irritation	0.05-2.0
Upper airway irritation	0.10-25
Lower airway and pulmonary effects	5.0-30
Pulmonary edema, inflammation, pneumonia	50.0-100
Death	100+

Table 2 Clinical Effects Irritation Index

Scale score	Description
10	Strong eye, nose, and throat irritation; great discomfort; strong odor
7	Moderate eye, nose, and throat irritation; discomfort
5	Mild eye, nose, and throat irritation; mild discomfort
3	Slight eye, nose, and throat irritation; slight discomfort
1	Minimal eye, nose, and throat irritation; minimal discomfort
0	No effects

Note. Derived from the literature by the Committee on Toxicology, "Formaldehyde—An Assessment of its Health Effects," NAS, 1980.

The degree of hypersensitivity to these responses in the population has not been defined.

The National Academy of Sciences Committee on Toxicology believes that the available controlled human studies are currently the most relevant sources for evaluation of the risks of formaldehyde in indoor air. These studies measured primary irritancy in test populations and provided dose-response data at various airborne concentrations of formaldehyde. These studies, however, were of short duration and obtained from limited test populations. When such data are applied to the general population, several factors may influence the extent of response, including variability of health status, genetic predisposition to the effects of irritants, and such physiological characteristics as age, sex, and pregnancy. In addition, responses reported in control studies may occur at an increased rate in the general population because of the interactions between formaldehyde and other irritants in the environment. Irritant effects of formaldehyde in humans are accelerated by the presence of cigarette smoke. The likelihood of interaction with other irritants, such as ozone and oxides of nitrogen, should not be discounted, but quantitative estimation of this combined effect is not now possible.

On the basis of consumer complaints and controlled human studies, irritation appears to be the most sensitive response to formaldehyde. The Committee on Toxicology's best judgment as to a range of irritation responses associated with exposure to various concentrations of formaldehyde is summarized in Table 2. This tabulation was developed from the limited number of controlled human studies, which provide the only dose-response data from human exposure to low airborne concentrations of formaldehyde.

The scoring scale shown here was evolved on an arbitrary basis after examining the several clinical studies with formaldehyde that produced dose-response effects (Table 3).

Although the extent of irritancy has not been investigated in controlled human studies at concentrations below 0.25 ppm, the Committee on Toxicology expects that fewer than 20 percent of an exposed human population would react

Table 3   Predicted Irritation Responses of Humans Exposed to Airborne
Formaldehyde

Concentration (ppm)	% of population giving indicated response	Degree of irritation[*]
1.5–3.0	20	7–10
	>30	5–7
0.5–1.5	10–20	5–7
	>30	3–5
0.25–0.5	20	3–5
<0.25	<20	1–3

[*]Irritation index: 10 = Strong irritation, great discomfort; 7 = Moderate irritation, discomfort; 5 = Mild irritation, mild discomfort; 3 = Slight irritation, slight discomfort; 1 = Minimal irritation, minimal discomfort; 0 = No effects.

to such formaldehyde exposure with slight irritation of the eyes, nose, and throat and possibly a slight decrease in nasal flow. As yet there is no evidence of a population threshold for the irritant effects of formaldehyde in humans (Table 3).

The preliminary results of an ongoing carcinogenicity study in rodents, reported in Chapter 11, the uncertainty about the variability of responses to formaldehyde in normal populations and in hypersensitive groups, and the current inadequacy of data (which leave unresolved the "no observed effect" dose in humans) all point to the advisability of maintaining formaldehyde at the lowest practical concentration to minimize adverse effects on public health.

REFERENCES

1   Altschuller, AP, IR Cohen, ME Meyer, and AF Wartburg, Jr. Analysis of aliphatic aldehydes in source effluents and in the atmosphere. *Anal. Chim. Acta* 25:101–117 (1961).

2   Kitchens, JF, RE Casner, GS Edwards, WE Harward III, and BJ Macri. Investigation of selected potential environmental contaminants: Formaldehyde. Final technical report. Atlantic Research Corp., Alexandria, VA. Rep. no. ARC 49-5681. EPA/560/2-76/009. 217 pp. (1976).

3   Stupfel, M. Recent advances in investigations of toxicity of automotive exhaust. *Environ. Health Perspect.* 17:253–285 (1976).

4   Health Effects of Formaldehyde. Final report to National Particleboard Association. Columbus, Ohio: Battelle Columbus Laboratories (1977b).

5   Kensler, CJ, and SP Battista. Components of cigarette smoke with ciliary-depressant activity: Their selective removal by filters containing activated charcoal granules. *N. Engl. J. Med.* 269:1161–1166 (1963).

6   Hollowell, CD, JV Berk, C-I Lin, and I Turiel. Indoor air quality in energy-efficient buildings. Lawrence Berkeley Laboratory, Energy and Environment

Division, University of California/Berkeley. (Report No.) LBL 8892, EEB Vent 79-2. 12 pp. (1979a).

7 U.S. Consumer Product Safety Commission. Ad Hoc Task Force, Epidemiology Study on Formaldehyde. Epidemiological studies in the context of assessment of the health impact of indoor air pollution. 34 pp. (1979).

8 Breysse, PA. Formaldehyde in mobile and conventional homes. *Environ. Health Safety News* 26(1-6). 20 pp. (1977).

9 Hollowell, CD, JV Berk, and GW Traynor. Impact of reduced infiltration and ventilation on indoor air quality in residential buildings. *ASHRAE Trans.* 85 (Part 1):816-826, 827 (1979b).

10 Akabane, J. Aldehydes and related compounds. In *International Encyclopedia of Pharmacological Therapy*, vol. 2, pp. 523-560 (1970).

11 Buss, J, K Kuschinsky, H Kewitz, and W Koransky. Enterale Resorption von Formaldehyd. Naunyn-Schmiedebergs Archiv. *Exp. Pathol. Pharmakol.* 247:380-381 (1964).

12 Williams, RT. *Detoxication Mechanisms: The Metabolism and Detoxication of Drugs, Toxic Substances and Other Organic Compounds*, 2nd ed., pp. 88-90. New York: Wiley & Sons (1959).

13 Goodman, JI, and TR Tephly. A comparison of rat and human liver formaldehyde dehydrogenase. *Biochim. Biophys. Acta* 252:489-505 (1971).

14 Strittmatter, P, and EG Ball. Formaldehyde dehydrogenase, a glutathione-dependent enzyme system. *J. Biol. Chem.* 213:445-461 (1955).

15 Neely, WB. The metabolic fate of formaldehyde-^{14}C intraperitoneally administered to the rat. *Biochem. Pharmacol.* 13:1137-1142 (1964).

16 Tephly, TR, WD Watkins, and JI Goodman. The biochemical toxicology of methanol. In WJ Hayes, Jr (ed.), *Essays in Toxicology*, vol. 5, pp. 149-177. New York: Academic Press (1974).

17 Uotila, L, and M. Koivusalo. Formaldehyde dehydrogenase from human liver: Purification, properties, and evidence for the formation of glutathione thiol esters by the enzyme. *J. Biol. Chem.* 249:7653-7663 (1974).

18 Doenecke, D. Digestion of chromosomal proteins in formaldehyde treated chromatin. *Hoppe-Seyler's Z. Physiol. Chem.* 359:1343-1352 (1978).

19 Shikama, K, and KI Miura. Equilibrium studies on the formaldehyde reaction with native DNA. *Eur. J. Biochem.* 63:39-46 (1976).

20 Berger, V, and M Tomas. Urea-formaldehyde resins with a low content of free formaldehyde. *Drevo* 19:211-214 (1964).

21 Harris, DK. Health problems in the manufacture and use of plastics. *Br. J. Ind. Med.* 10:255-268 (1953).

22 Wild, H. The liberation of formaldehyde during the hardening of urea-formaldehyde resins. *Holztechnot.* 5:92-95 (1964). [English abstract in *Abstr. Bull. Inst. Paper Chem.* 35:1120-1121 (1965).]

23 U.S. Department of Health, Education, and Welfare, Public Health Service, Center for Disease Control, National Institute for Occupational Safety and Health. *Criteria for a Recommended Standard . . . Occupational Exposure to Formaldehyde.* DHEW (NIOSH) Publication No. 77-126. Washington, D.C.: Government Printing Office (1976).

24 Andersen, I. Formaldehyde in the indoor environment—health implications

and the setting of standards. In PO Fanger, and O Valbjorn (eds.), *Indoor Climate: Effects on Human Comfort, Performance and Health in Residential, Commercial, and Light-Industry Buildings*, pp. 65–77, 77–87. Proceedings of the First International Indoor Climate Symposium, Copenhagen, August 30–September 1, 1978. Copenhagen: Danish Building Research Institute (1979).

25 Weber-Tschopp, AT Fischer, and E Grandjean. Reizwirkungen des Formaldehyds (HCHO) auf den Menschen (Irritating effects of formaldehyde on men). *Int. Arch. Occup. Environ. Health* 39:207–218 (1977).

26 Schuck, EA, ER Stephens, and JT Middleton. Eye irritation response at low concentrations of irritants. *Arch. Environ. Health* 13:570–575 (1966).

27 Glass, WI. An outbreak of formaldehyde dermatitis. *N.Z. Med. J.* 60:423–427 (1961).

28 Pirila, V, and O Kilpio. On dermatitis caused by formaldehyde and its compounds. *Ann. Med. Intern. Fenn.* 38:38–51 (1949).

29 Rostenberg, A, Jr, B Bairstow, and TW Luther. A study of eczematous sensitivity to formaldehyde. *J. Invest. Dermatol.* 19:459–462 (1952).

30 Blejer, HP, and BH Miller. Occupational health report of formaldehyde concentrations and effects on workers at the Bayly Manufacturing Company, Visalia, Calif. Study report number S-1806. Los Angeles, State of California Health and Welfare Agency, Dept. of Public Health, Bureau of Occupational Health (1966).

31 Marzulli, FN, and HI Maibach. Antimicrobials: Experimental contact sensitization in man. *J. Soc. Cosmet. Chem.* 24:399–421 (1973).

32 Jordan, WP, Jr, WT Sherman, and SE King. Threshold responses in formaldehyde-sensitive subjects. *J. Am. Acad. Dermatol.* 1:44–48 (1979).

33 Maibach, HI, and T Franz. Provocative use tests with formalin: Dose-response relationships (this volume).

34 Morrill, EE, Jr. Formaldehyde exposure from paper process solved by air sampling and current studies. *Air Cond. Heat. Vent.* 58(7):94–95 (1961).

35 Ettinger, I, and M Jeremias. A study of the health hazards involved in working with flameproofed fabrics. New York State Dept. Labor, Div. Ind. Hyg., *Mon. Rev.* 34:25–27 (1955).

36 Zannini, D, and L Russo. Long-standing lesions in the respiratory tract following acute poisoning with irritating gases. *Lav. Um.* 9:241–254 (1957).

Chapter 2

Introduction to Formaldehyde Toxicity

Richard A. Griesemer

In Chapter 1, Newell set the stage for consideration of the toxicity of formaldehyde. In the succeeding chapters we will review the available experimental laboratory data and attempt to assess *a*) the toxicity of formaldehyde relative to other toxic chemicals, *b*) the range of animal species that are similarly affected, *c*) the kinds of disease produced, *d*) the parts of the body affected, and *e*) the effects of potential host-modifying factors such as age or pregnancy. Since the acute toxic effects of formaldehyde have recently been reviewed by the National Academy of Sciences (1), we will concentrate instead on the chronic effects from repeated exposures.

The cornerstone for this presentation is the CIIT bioassay of formaldehyde by inhalation in rats and mice. Conceptually, there are few reasons to believe that the approach to evaluation of formaldehyde toxicity should differ from that for other chemicals. One possible difference stems from formaldehyde's irritating and cytotoxic properties, which, as for other chemicals that are acutely

The U.S. Government's right to retain a nonexclusive, royalty-free license in and to the copyright covering this paper, for governmental purposes, is acknowledged.

toxic, limit the dose levels that can be tested. Although testing at relatively low doses reduces the likelihood of detecting adverse effects, it provides the possibility of detecting effects at dose levels comparable to those to which humans may be exposed. A possible problem in the toxicological evaluation is the identification or characterization of what was tested in each analysis. Was part of the formaldehyde in a linear or cyclic polymeric form? How much methanol and other additives or impurities did the test material contain? Perhaps data on the toxicity of other aldehydes will contribute to our understanding of formaldehyde toxicity. It may be instructive to note that a closely related aldehyde, acetaldehyde, was recently reported to cause nasal and laryngeal cancers in hamsters (2) and that 16 aldehydes, including formaldehyde and acetaldehyde, have been reported to induce tumors in laboratory animals (mostly local tumors following topical or subcutaneous administration) (National Cancer Institute, unpublished data).

In the toxicological evaluation of formaldehyde, there are some additional considerations. In measuring the doses delivered by various routes of exposure, it would be helpful (where data exist) to know the doses delivered to the target tissue. Does formaldehyde reach the germinal cells? Do the pathways and rates of metabolism differ in different species or in different tissues? What is known about the toxicity of the metabolites, including formate? What is the magnitude and significance of the endogenous formation of formaldehyde? Is an adequately designed and conducted negative teratology study in the mouse conclusive, or must other species be studied? To what extent to in vitro cell transformation assays support in vivo carcinogenicity assays? Does in vitro transformation of mouse fibroblastic cells, for example, indicate that the test substance is likely to produce sarcomas in mice? How does one interpret mutagenicity assays that are positive but so weak that dose-response relationships are difficult to establish? What is the significance of the reported synergistic mutagenic effects of formaldehyde with X-radiation, ultraviolet light, and peroxide?

Regarding the CIIT bioassay, are there any other experimental data to support the CIIT results? Is the carcinogenicity of formaldehyde restricted to one species (the rat) and one tissue (the nose)? Why is the nasal mucosa seemingly the only target tissue in rats for the carcinogenicity of inhaled formaldehyde? Was sialodacryoadenitis a significant factor in nasal tumor development in the CIIT study? Do the irritant properties of formaldehyde contribute to its carcinogenicity?

It is hoped that the answers to these questions and others will assist in critically weighing the evidence for the toxicity of formaldehyde.

REFERENCES

1 National Academy of Sciences. *Formaldehyde—An Assessment of Its Health Effects*. Washington, D.C.: Committee on Toxicology, National Academy of Sciences (1980).

2   Feron, VJ, and RA Woutensen. Respiratory tract tumors in hamsters ex-
    posed to acetaldehyde vapour alone or simultaneously to benzo(a)pyrene.
    Abstract for the Symposium on Cocarcinogenesis and Biological Effects
    of Tumor Promoters (October 13-16), Elmau Castle, Klais, Bavaria, Ger-
    many, sponsored by the German Cancer Research Center, Heidelberg, and
    IARC, Lyon, France (1980).

Chapter 3

Formaldehyde Sensory Irritation

Craig S. Barrow

William H. Steinhagen

Jane C. F. Chang

The membranes of the respiratory tract are endowed with a wide variety of sensory nerve endings that are capable of responding to chemical and/or physical stimuli. One such group of nerve endings is associated with the maxillary and ophthalmic divisions of the trigeminal nerve in the nasal mucosa. Stimulation of these by airborne chemical irritants (e.g., formaldehyde) results in a painful burning sensation, a desire to withdraw from the contaminated atmosphere, and a decrease in respiratory rate (1). In larger mammals, including humans, stimulation of trigeminal nerve endings also evokes the sneeze reflex (2, 3). This series of responses has been termed the "common chemical sense," separating it from more specialized chemical senses such as olfaction or gustation (4).

The term "sensory irritation" is synonymous with the common chemical sense and has been used to describe one effect of chemicals which, when inhaled via the nose, stimulate trigeminal nerve endings, evoke a burning sensation of the nasal passages, and inhibit respiration (1). All substances that may excite the

The authors would like to thank Drs. J. Swenberg and J. Popp and Ms. E. Gross of the CIIT Pathology Department for their elegant morphometric studies of the rat and mouse nasal cavities.

common chemical sense are potentially noxious and lung damaging. These reflex actions represent an important respiratory tract defense mechanism by serving to minimize inhalation of the noxious agent and by warning of its presence through the perception of pain (1, 5, 6). The cornea is also innervated by the trigeminal nerve (7) and exposure to airborne chemical irritants can cause intense burning and lacrimation.

Together with the decrease in respiratory rate is a series of other physiological responses that may occur during exposure to sensory irritants (8). The mechanisms that compensate for the decreased pulmonary ventilation include peripheral vasoconstriction, secretion of catecholamines, and bradycardia (8). The decrease in respiratory rate, however, remains as the principal event, and measurements of the decrease in respiratory rate during exposure to a chemical can be used as a quantitative measure for evaluation of sensory irritation (1). A concentration-response curve can be obtained by plotting the logarithm of the exposure concentration versus the maximum percent decrease from control values in respiratory rate. Concentration-response curves have been published for a wide variety of sensory irritants, including formaldehyde (HCHO)[*] (9, 10).

The sensory irritation properties of HCHO are well known. In humans, irritation of the eyes, nose, and throat has been reported to occur below 1 ppm (11–13). The sensory irritation response in Swiss-Webster mice after single or repeated HCHO exposures (9) and in combination with acrolein (14) has been reported. Studies have also been published on the electrical activity of the nasopalantine nerve (maxillary division of trigeminal) and ethmoidal nerve (ophthalmic division of trigeminal) of anesthetized Sprague-Dawley rats exposed to HCHO (15).

Studies in this laboratory have been conducted in order to further understand the sensory irritation properties of HCHO. Our objectives were:

1 To investigate species differences between male F-344 rats and male B6C3F1 mice

2 To investigate the effects of single versus repeated inhalation of HCHO on the sensory irritation response in these species

3 To determine to what extent tidal volume changes compensate for the decrease in respiratory rate during exposure to HCHO

MATERIALS AND METHODS

Animals, Exposure Concentrations, and Measurement of Endpoint Response

Male F-344 rats (Fischer-344, CDF (F-344)/Crl BR, Charles River Breeding Labs, Kingston, NY) weighing 170–225 g were divided into four groups (24–30 rats/group). Effects of HCHO exposure on respiratory rate and tidal volume were studied in naive (no previous exposure to HCHO) or pretreated rats (HCHO, 15

---

[*]HCHO and CH_2O are used interchangeably throughout the text.

ppm, 6 h/day for 4 days). Additionally, male B6C3F1 mice ($B_6C_3F_1$/Crl BR, Charles River Breeding Labs, Kingston, NY) weighing 20–30 g were divided into four groups (23–28 mice/group) for similar studies of respiratory rate and tidal volume. Methods used to quantitate respiratory rate (1, 8, 9, 10) and tidal volume (16) have been previously described.

Concentration-response curves for either respiratory rate or tidal volume were obtained by exposing groups of three to four animals for 10 min in head-only chambers. The range of HCHO concentrations studied was approximately 1–50 ppm for rats and 0.5–15 ppm for mice. Head-only exposures of pretreated rats or mice were conducted 18–24 h following the last day of pretreatment.

Exposure Chambers

Whole-body exposures of rats or mice were conducted using dynamic airflow ($\geqslant 100$ liters/min) in a 391-liter all glass exposure chamber fabricated from a glass aquarium (17). Respiratory rate or tidal volume measurements were made in a head-only exposure chamber (18) during exposure to HCHO. Each chamber was equipped with four plethysmographs. Chamber volumes were 6.11 liters for rats and 2.67 liters for mice (18). Airflow in these chambers was > 20 liters/min.

Generation and Analysis of Test Atmospheres

During whole-body exposures, generation of the HCHO test atmosphere was accomplished by vaporizing paraformaldehyde (Aldrich) at a controlled rate in a thermostatically controlled oven maintained at isothermal conditions. A three-way valve enabled the stream of HCHO gas to be vented to a hood or to the chamber supply inlet. The target concentration was obtained by altering the exhaust airflow through the chamber (100–200 liters/min).

For head-only exposures, a known amount of paraformaldehyde (Aldrich) was vaporized in a specially constructed flask and passed into a 50-liter Teflon bag with Ultra Zero Air (Matheson Gas). The final concentration of HCHO in the bag ranged from 2500 to 3000 ppm. Target concentrations were achieved by delivering HCHO at a constant flow rate via a peristaltic pump into the inlet of the head-only chamber. Final concentrations were achieved by varying the HCHO flow rate or exhaust air through the exposure chamber (20–175 liters/min).

Analysis of HCHO was accomplished by infrared spectrophotometry at a wavelength of 3.58 μm and a path length of 20.25 meters. The infrared analyzer was calibrated with a paraformaldehyde permeation tube. The permeation rate was quantitated by a chromotropic acid procedure (19).

RESULTS

Exposure Chamber Data

The mean time-weighted average for all 4-day pretreatments was 14.7 ppm (±0.5, 13.7–15.8; standard deviation and range).

HCHO Effects on Respiratory Rate

A concentration-dependent decrease in respiratory rate was seen in both naive and pretreated rats (Fig. 1) during HCHO inhalation exposure. However, tolerance development was seen after 4 min in both groups, as evidenced by the return of respiratory rate toward control values. This was more apparent at concentrations above 4 ppm.

In contrast to these results, naive or pretreated mice did not develop tolerance to HCHO (Fig. 1). Exposures were characterized by an abrupt decrease in

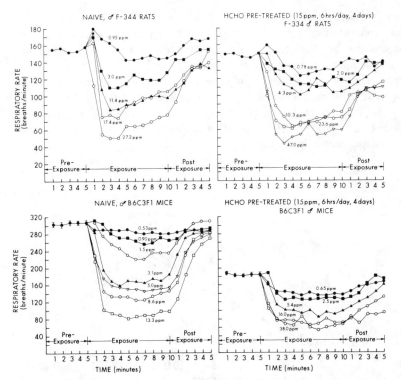

Figure 1 Time-response curves of respiratory rate obtained with naive or HCHO-pretreated rats and mice exposed for 10 min to various concentrations of HCHO.

respiratory rate, which plateaued in 3-4 min and persisted throughout the 10-min exposures. Preexposure respiratory rate baselines were significantly different between the two groups ($\bar{x}$ naive mice = 305 breaths/min vs $\bar{x}$ pretreated mice = 184 breaths/min).

Concentration-response curves from rats based on maximum percent decrease in respiratory rate were very similar for naive or pretreated animals (Fig. 2). The RD_{50} values (concentration expected to elicit a 50 percent decrease in respiratory rate) with 95 percent confidence intervals were similar for both groups of rats (naive = 13.1 ppm, 10.6-17.5 ppm; HCHO pretreated = 10.8 ppm, 7.6-16.9 ppm). An examination of concentration-response curves from mice showed very similar RD_{50} values (naive = 4.4 ppm, 0.9-5.0 ppm; pretreated = 4.3, 3.4-5.5 ppm). However, as shown in Fig. 3, the slopes from the regression lines were significantly different ($p \leqslant 0.01$).

HCHO Effects on Tidal Volume

Exposure of naive or pretreated rats resulted in increased tidal volume (Fig. 4). This was concentration dependent only in the pretreated animals. Also important was the fact that baseline tidal volume was higher (~ 1.38 ml) in pretreated rats compared to naive rats (~ 1.1 ml).

Comparison of tidal volume from naive and pretreated mice showed a slight increase with naive animals but a decrease in pretreated ones (Fig. 4). Baseline tidal volume of pretreated mice (0.28 ml) was 35 percent higher compared to naive mice (0.207 ml). The effect of HCHO exposure on tidal volume was concentration dependent in both groups.

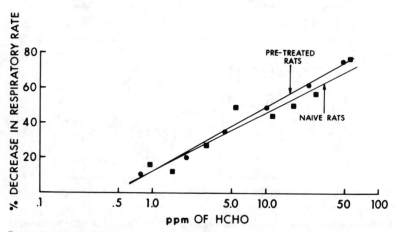

Figure 2 HCHO concentration-response curves from naive or pretreated (15 ppm HCHO 6 h/day, 4 days) male F-344 rats. The maximum percentage decrease in respiratory rate that occurred during a 10-min exposure was plotted for each concentration.

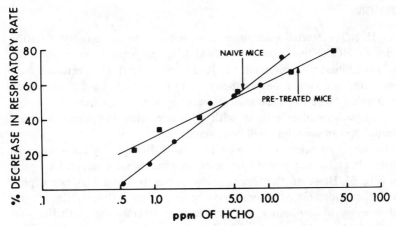

Figure 3 HCHO concentration-response curves from naive or pretreated (15 ppm HCHO 6 h/day, 4 days) male B6C3F1 mice. The maximum percentage decrease in respiratory rate that occurred during a 10-min exposure was plotted for each concentration.

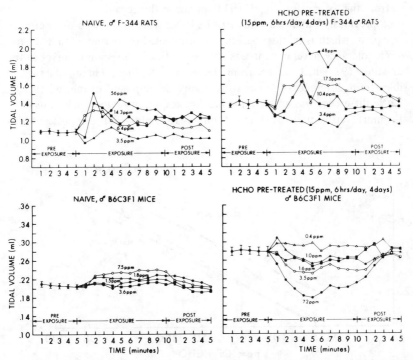

Figure 4 Time-response curves of tidal volume obtained with naive or HCHO-pretreated rats and mice exposed for 10 min to various concentrations of HCHO.

DISCUSSION

Naive or HCHO-pretreated mice were more sensitive to the sensory irritation properties of HCHO than rats (Fig. 5). At RD_{50} this was approximately 0.5 order of magnitude (~4 ppm in mice vs 10-13 ppm in rats). The decreased slope with pretreated mice was probably the result of the reduced baseline in the pre-treated animals. Since the maximum decrease appears to be fixed at approximately 60 breaths/min, in a group with a lower baseline the percent decrease will be much less and the slope will be lowered.

Pretreatment of mice or rats with 15 ppm HCHO 6 h/day for 4 days had very little effect on the sensory irritation response when compared to naive animals (Fig. 5). However, the time-response curves for respiratory rate depression showed considerable differences between rats and mice (Fig. 1). Rats clearly showed recovery of respiratory rate or tolerance development to HCHO after approximately 4 min of exposure. This became more pronounced at concentrations above 4 ppm. Mice exposed to HCHO did not show any indications of recovery of respiratory rate during the 10-min HCHO exposures (Fig. 3). This finding is relevant since tolerance development could result in higher concentrations of HCHO penetrating the respiratory tract. Mice appear to be better protected from the toxic effects of HCHO than rats in this regard.

In a previously published study (9) the RD_{50} from male Swiss-Webster mice was 3.1 ppm, which is in close agreement with male B6C3F1 mice studied here. However, tolerance development was seen from the time-response curves in the earlier study (9), which differs from the present findings. These data suggest possible strain differences in mice to tolerance development of inhaled HCHO.

Little difference was noted in the baseline respiratory rate of naive rats (155 breaths/min) versus pretreated rats (148 breaths/min), but a significant dif-

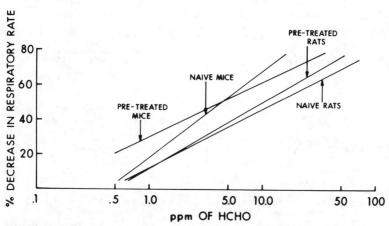

Figure 5 Comparison of HCHO concentration-response curves between naive or HCHO-pretreated B6C3F1 mice versus F-344 rats.

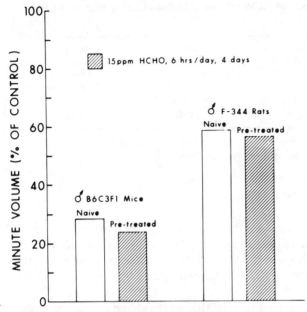

Figure 6 Effects of HCHO exposure (15 ppm, 10 min) on minute volume in rats and mice. Minute volume changes were calculated from the respective respiratory rate and tidal volume concentration-response curves for each group of animals.

ference was observed between mouse groups (naive = 305 breaths/min vs 184 breaths/min with pretreated mice). This represents a 40 percent decrease in control respiratory rate. The mechanism for this effect is unknown.

Baseline tidal volume was increased by over 35 percent in pretreated mice to compensate for the 40 percent decrease in rate. Rats also showed this effect even though the baseline respiratory rate was similar in both groups. During exposure, the increased tidal volume did not compensate entirely for the decrease in rate. For example, at the RD_{50} value in rats the maximum tidal volume increase was approximately 21 percent in naive animals and 19 percent in pretreated rats. For naive mice, the increase was 14 percent while in pretreated mice tidal volume decreased by 19 percent. These results indicate that tidal volume does not compensate entirely for the decreased rate and that the compensation is slightly greater in rats than in mice. As a result, minute volume (respiratory rate × tidal volume) decreased during exposure to HCHO.

The relevance of the decreased minute volume can be shown by comparing all groups at 15 ppm (Fig. 6). Mice were able to decrease their minute volume by approximately 75 percent as compared to 45 percent in rats. These results indicate that the respiratory tract of the B6C3F1 mouse may be better protected than that of the F-344 rat to inhaled HCHO. Little difference was seen in naive

Table 1   Theoretical Dose[*] of HCHO to Nasal Mucosa During 15 ppm HCHO Exposure

	Rat[†]	Mouse[†]
HCHO conc. (μg/liter)	18.4	18.4
Minute volume (liters/min)	0.114	0.012
Nasal cavity surface area (cm²)	13.44	2.89
"Dose" (μg/min/cm²)	0.156	0.076

$$*"Dose" = \frac{\text{HCHO conc. } (\mu g/liter) \times \text{minute volume (liters/min)}}{\text{nasal cavity surface area (cm}^2)}$$

$$= \mu g/min/cm^2$$

[†]Pretreated to 15 ppm HCHO, 6 h/day, 4 days.

vs pretreated animals. It should also be pointed out that Fig. 6 actually depicts the minimum percent of control in minute volume and does not take into account the tolerance development seen from the time-response curves in rats. Considering this the differences in total exposure would be even greater than in Fig. 6.

Due to the high water solubility of HCHO and the obligatory nose breathing characteristics of rodents, the nasal cavity is considered to be a primary target organ. From these data an approximation can be made of the dose of HCHO to the nasal mucosa in these two species. If dose is expressed as μg/min/cm² (Table 1) for a 10-min 15 ppm HCHO exposure in the pretreated animals, a two-fold increase in "dose" would be expected in rats versus mice. This is consistent with the increased frequency of squamous cell carcinoma of the nasal cavity seen in F-344 rats when compared to B6C3F1 mice exposed to 15 ppm HCHO (20). Further support for this hypothesis is necessary, particularly longer exposure times, i.e., 6 h. Other mechanisms that may influence species differences to inhaled HCHO toxicity include metabolism, mucociliary clearance, and nasal cavity blood flow. Nevertheless, the respiratory tract reflexes described here may be effective in protecting the respiratory tract from the toxic effects of inhaled HCHO.

REFERENCES

1   Alarie, Y. Sensory irritation by airborne chemicals. *CRC Crit. Rev. Toxicol.* 2:299–363 (1973).

2   Angell James, JE, and MdeB Daly. Nasal reflexes. *Proc. R. Soc. Med.* 62: 1287–1293 (1969).

3   Widdicombe, JG. Reflex control of breathing. In *Respiratory Physiology*, edited by JG Widdicombe, vol. 2, pp. 273–301. Baltimore: Univ. Park Press (1974).

4 Keele, CA. The common chemical sense and its receptors. *Arch. Int. Pharmacodyn. Ther.* 139:547–557 (1962).

5 Widdicombe, JG. Defense mechanisms of the respiratory tract and lungs. In *Respiratory Physiology II*, edited by JG Widdicombe, vol. 14, pp. 291–315. Baltimore: Univ. Park Press (1977).

6 Comroe, JH, Jr. Defense mechanisms of the lungs. In *Physiology of Respiration*, 2d ed. pp. 220–228. Chicago: Year Book (1974).

7 Whitear, M. An electron microscope study of the cornea in mice, with special reference to the innervation. *J. Anta.* 94:387–409 (1960).

8 Barrow, CS, Y Alarie, and MF Stock. Sensory irritation and incapacitation evoked by thermal decomposition products of polymers and comparisons with known sensory irritants. *Arch. Environ. Health* 33(2):79–88 (1978).

9 Kane, LE, and Y Alarie. Sensory irritation to formaldehyde and acrolein during single and repeated exposures in mice. *Am. Ind. Hyg. Assoc. J.* 38(10):509–522 (1977).

10 Kane, LE, CS Barrow, and Y Alarie. A short-term test to predict acceptable levels of exposure to airborne sensory irritants. *Am. Ind. Hyg. Assoc. J.* 40(3):207–229 (1979).

11 Bourne, HG, Jr, and S Seferian. Formaldehyde in wrinkle-proof apparel processes—tears for milady. *Ind. Med. Surg.* 28:232–233 (1959).

12 Kerfoot, EJ, and TF Mooney, Jr. Formaldehyde and paraformaldehyde study in funeral homes. *Am. Ind. Hyg. Assoc. J.* 36:533–537 (1975).

13 Weber-Tschopp, A, T Fischer, and E. Grandjean. Irritating effects of formaldehyde (HCHO) on humans. *Int. Arch. Occup. Environ. Health* 39:207–218 (1977).

14 Kane, LE, and Y Alarie. Evaluation of sensory irritation from acrolein-formaldehyde mixtures. *Am. Ind. Hyg. Assoc. J.* 39:270–274 (1978).

15 Kulle, TJ, and GP Cooper. Effects of formaldehyde and ozone on the trigeminal nasal sensory system. *Arch. Environ. Health* 30:237–243 (1975).

16 Barrow, CS, and WH Steinhagen. NH_3 concentrations in the expired air of the rat: Importance to inhalation toxicology. *Toxicol. Appl. Pharmacol.* 53:116–121 (1980).

17 Barrow, CS, and DE Dodd. Ammonia production in inhalation chambers and its relevance to chlorine inhalation studies. *Toxicol. Appl. Pharmacol.* 49:89–95 (1979).

18 Barrow, CS, Y Alarie, JC Warrick, and MF Stock. Comparison of the sensory irritation response in mice to chlorine and hydrogen chloride. *Arch. Environ. Health* 32(2):68–76 (1977).

19 Katz, M. Tentative method of analysis for formaldehyde content of the atmosphere (colorimetric method). In *Methods of Air Sampling and Analysis*, edited by M Katz, 2d ed., pp. 303–307. Washington, DC: American Public Health Assoc. (1977).

20 Swenberg, JA, WD Kerns, RI Mitchell, EJ Gralla, and KL Pavkov. Induction of squamous cell carcinomas of the rat nasal cavity by inhalation exposure to formaldehyde vapor. *Cancer Res.* 40:3398–3402 (1980).

Chapter 4

Distribution of [^{14}C] Formaldehyde in Rats after Inhalation Exposure

Henry d'A. Heck

Theodore Y. Chin

Mercedes C. Schmitz

The disposition of formaldehyde following oral, intravenous, or intraperitoneal administration has been studied in several animal species (1-8). No investigations of CH_2O disposition following inhalation exposure have, however, been reported. Since inhalation of moderate to high concentrations of airborne CH_2O by rats has been correlated with the development of nasal carcinomas (9), and because inhalation is the major route of human exposure to CH_2O, studies of the distribution of CH_2O were begun in rats using this exposure method.

The principal goal of these experiments is to determine the relationship between airborne CH_2O concentrations and the corresponding tissue concentrations and to relate the latter to the induction of neoplasia. Particular attention has, therefore, been focused on the upper respiratory tract, especially the nasal turbinates, where the carcinomas have been shown to develop. In addition, there is a slight possibility that airborne CH_2O exposure could result in toxic concentrations of CH_2O or of CH_2O metabolites being attained in other tissues. There-

We thank Dr. Craig S. Barrow for carrying out the preexposures of rats to airborne CH_2O, Dr. James S. Bus for the use of his head-only inhalation chamber, and Dr. James A. Popp for assistance in the development of the mucosal isolation method.

fore, the distribution of CH_2O and its metabolites throughout the animal will also be described.

In our experiments, Fischer-344 (F-344) rats were exposed in a head-only chamber to different concentrations of airborne CH_2O labeled with carbon-14. Exposure of only the head, rather than the entire body, simplifies the interpretation of the results, since absorption through the skin, if it occurs, is limited to a relatively small surface area. Radioactivity distributed to the tissues was determined by scintillation counting. The concentrations of radioactivity in the tissues are usually expressed in this paper as μmole or nmole equivalents of $^{14}CH_2O$ per gram of tissue to provide a common basis for comparison among different tissues or fluids. It should be emphasized, however, that due to metabolism, the radioactivity may, in fact, derive from a number of different chemical species.

MATERIALS AND METHODS

Chemicals

$[^{14}C]$paraformaldehyde was purchased from Amersham. Gaseous $^{14}CH_2O$ was prepared by heating the $[^{14}C]$paraformaldehyde together with unlabeled paraformaldehyde (Aldrich) in a modified filtering flask, while simultaneously passing dry compressed air through the flask and into a 60-liter Teflon bag. The concentration of formaldehyde in the bag was determined on aliquots analyzed by either the chromotropic acid method (10) or an infrared absorption technique (Wilks MIRAN-1A). Specific activities were calculated by adding known volumes from the bag to water in sealed serum bottles, followed by scintillation counting of the solutions. Aqueous $[^{14}C]$formaldehyde and aqueous sodium $[^{14}C]$formate for injection were purchased from Amersham and New England Nuclear.

Animals

Male F-344 rats (Charles River, Portage, MI), weighing between 180 and 250 g, were used for all exposures. Rats received NIH-07 laboratory diet (Zeigler Brothers, Gardners, PA) and tap water ad libitum. The rats were held in ventilated restrainers during the exposure period (6 h) with only their heads protruding into the exposure chamber. Room temperature was maintained at 70–75°F.

Exposure Procedure

All exposures were carried out under dynamic conditions. A peristaltic pump was used to meter $[^{14}C]$formaldehyde from the bag into a three-necked flask, where the compound was diluted with room air. The diluted gas was passed through one neck of the flask into the exposure chamber, thence over charcoal and molecular sieve traps, using a pump located downstream from these components. The concentration of formaldehyde in the chamber was determined either by measuring the radioactivity of withdrawn aliquots or by continuous monitoring with the infrared spectrophotometer.

To examine the effects of preexposure to formaldehyde on $^{14}CH_2O$ tissue concentrations, four rats were exposed (6 h/day for 9 days) to unlabeled CH_2O at 15 ppm in a whole-body exposure chamber. On the tenth day, the rats were exposed in a head-only chamber to 14.9 ppm of $^{14}CH_2O$ for 6 h. Simultaneously, a second group of four rats that had not previously been exposed to CH_2O were exposed to the same $^{14}CH_2O$ vapor. All rats were sacrificed immediately after exposure and tissue concentrations of radioactivity were determined.

Tissue Distribution of Radioactivity

Rats were anesthetized with methoxyflurane (Metofane) immediately after exposure, the thoracic cavity was opened, and the animals were exsanguinated by cardiac puncture. Tissues [liver, kidneys, spleen, small intestine (approximately 20 cm from the central section), testes, lungs, heart, esophagus, and trachea], as well as the head, were removed immediately and stored in ice. Blood was centrifuged, and the plasma and packed cells were separated. All samples were frozen for subsequent analysis.

Tissues from the head were collected in the following manner. The brain was removed and saved, and the mandible was discarded. The dorsal bone covering the nasal cavity and the hard palate lying beneath were split to provide access to the nasal turbinates. Mucosal tissue was scraped from the nasal turbinates, maxilloturbinates, and lateral walls of the nasal cavity (11). In the CIIT-sponsored long-term toxicity study of CH_2O, it was usually the tissue in these regions of the nasal cavity that exhibited a neoplastic response (9). Mucosa was also collected from the median septum (11). This tissue did not exhibit a neoplastic response in the long-term study, but since it is adjacent to the tissues that did respond, it was of interest to examine this tissue as well. In the normal rat, both regions of mucosal tissue that were collected were composed of respiratory epithelium (11). Concentrations of radioactivity in the two mucosal regions were determined separately.

After homogenization in water, tissue samples were digested for approximately 16 h in Protosol/C_2H_5OH (1/2) (New England Nuclear) at 55°C. The samples were decolorized by incubation for an additional 30 min in the presence of H_2O_2, neutralized with HCl, and diluted in Amersham Aqueous Counting Scintillant for analysis of radioactivity.

Metabolism and Elimination of ^{14}CH$_2$O

Immediately after a 6-h exposure, rats were transferred to individual metabolism cages, which permitted continuous collection of expired air, urine, and feces. The rats remained in the cages for 70 h, during which time access to food and water was unrestricted. Exhaled ^{14}CO$_2$ was trapped in a solution of 5 M ethanolamine in 2-methoxyethanol. The traps were changed several times to permit determinations of ^{14}CO$_2$ excretion rates. Urine and feces were frozen in dry ice throughout the period of collection. At the end of this period, rat tissues were removed and frozen as described above. Carcasses were saved for an estimation of total residual radioactivity. Metabolism cages were washed and the washings combined with the urine collections.

Pharmacokinetics

The uptake and disappearance of radioactivity in blood following inhalation exposure were determined in rats using a cannulation method. The cannula was installed in the jugular vein and exited dorsally in the region of the scapulae, which allowed blood samples to be drawn both during and after inhalation exposure. The investigations of pharmacokinetics following a bolus intravenous injection of [^{14}C]formaldehyde or of [^{14}C]formate were carried out similarly, with injection occurring into the tail vein. Concentrations of total radioactivity in whole blood and in plasma were determined in each blood sample (approximately 0.2 ml). Hematocrits (Hct) were measured, and the concentration of radioactivity in the packed cell fraction was calculated using the equation,

$$[^{14}C]\,packed\,cells = \{[^{14}C]\,whole\,blood - (1 - Hct)[^{14}C]\,plasma\}/Hct$$

RESULTS

Absorption and Distribution of Airborne ^{14}CH$_2$O

Determinations of total radioactivity concentrations in the nasal mucosa, trachea, and plasma immediately after a 6-h exposure are summarized in Fig. 1. Concentrations were much higher in the nasal mucosa than in any other tissue examined. Absorption appears, therefore, to occur primarily in the upper respiratory tract, as would be expected on the basis of the high aqueous solubility of formaldehyde.

The concentrations of radioactivity in the mucosal linings of the nasal turbinates, maxilloturbinates, and lateral walls did not differ significantly from the concentrations measured in the mucosal linings of the median septum. Consequently, the tissue concentrations in both mucosal regions were combined, and the average concentrations are reported in Fig. 1.

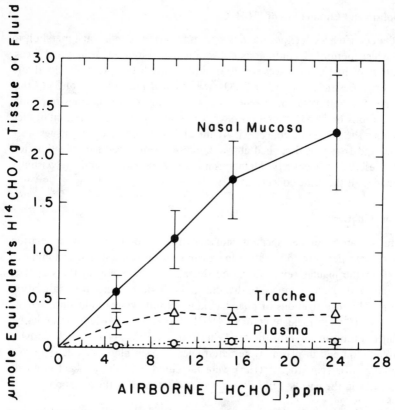

Figure 1 Equivalent concentrations of [^{14}C] formaldehyde in the nasal mucosa, trachea, and plasma of male F-344 rats immediately after a 6-h exposure to selected concentrations of airborne [^{14}C] formaldehyde. Each value was the average from three exposures, each exposure involving four rats. Error bars represent overall standard deviations.

The effects of preexposure to 15 ppm of formaldehyde on the equivalent concentrations of $^{14}CH_2O$ in the nasal mucosa and plasma of rats after a 6-h exposure to 15 ppm of $^{14}CH_2O$ are shown in Table 1. There were no detectable differences in the equivalent tissue concentrations of $^{14}CH_2O$ in naive or preexposed animals. Thus, preexposure for 9 days to 15 ppm of formaldehyde vapor does not appear to influence either the absorption of this chemical in the nasal turbinates or its distribution to plasma.

In contrast to the nasal mucosa and trachea, both of which are exposed directly to airborne CH_2O, exposure of the internal organs and tissues to exogenous formaldehyde must occur after absorption. To estimate the maximum level of potential exposure, tissue radioactivity concentrations were determined. In all cases, the equivalent tissue concentrations were very low, resembling the plasma concentrations. The ratios of radioactivities in the different tissues to

Table 1   Effects of Preexposure of F-344 Rats to 15 ppm Formaldehyde (6 h/day for 9 days) on Equivalent Concentrations of ^{14}CH$_2$O in Nasal Mucosa and Plasma Immediately After a 6-h Exposure to 14.9 ppm Airborne ^{14}CH$_2$O on Day 10

Tissue	Equivalent [^{14}CH$_2$O], nmoles/g		Significance level[†]
	Naive rats[*]	Preexposed rats[*]	
Nasal mucosa	2148 ± 255	2251 ± 306	NS ($p > 0.5$)
Plasma	76 ± 11	79 ± 7	NS ($p > 0.5$)

[*]Mean ± SD. Four rats per exposure group.
[†]Two-tailed t-test. NS = not significant.

that in plasma appeared to be essentially constant, independent of the airborne concentration. Average ratios for all of the airborne concentrations are presented in Table 2.

With the exception of the esophagus, the concentrations of radioactivity in the internal organs and tissues of the rat differed by no more than about three-fold from those observed in plasma (Table 2). The relatively high concentrations found in the esophagus may be a result of mucociliary action in the nasal cavity, the trachea, or both. It should be remembered, however, that these ratios are based on total radioactivity and do not necessarily reflect the levels of exogenous formaldehyde in the different tissues.

Table 2   Concentration of Radioactivity in Tissues Relative to That in Plasma Immediately After Exposure of F-344 Rats to Airborne [^{14}C] Formaldehyde for 6 h

Tissue	(DPM/g tissue)/(DPM/g plasma)[*]
Esophagus	4.94 ± 1.23
Kidney	3.12 ± 0.47
Liver	2.77 ± 0.25
Intestine	2.64 ± 0.48
Lung	2.05 ± 0.36
Spleen	1.59 ± 0.50
Heart	1.09 ± 0.09
Brain	0.37 ± 0.06
Testes	0.31 ± 0.05
Erythrocytes	0.30 ± 0.08

[*]Mean ± SD. Ratios were calculated at each of four exposure concentrations (range: 5–24 ppm, 12 rats at each concentration) and were averaged. Observed ratios appeared to be independent of the exposure concentration.

Elimination of Airborne $^{14}CH_2O$

Following a 6-h exposure to either 0.63 or 13.1 ppm of airborne $^{14}CH_2O$, excretion of radioactivity was determined in the expired air, urine, and feces over a period of 70 h. The average $^{14}CO_2$ excretion rates, plotted in Fig. 2, are multiphasic, involving an initial, relatively high rate of excretion, which declines

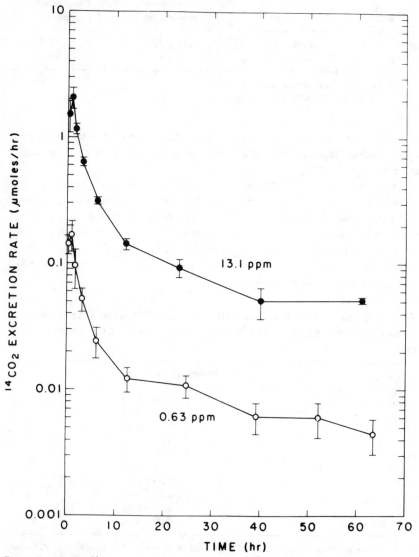

Figure 2 Average $^{14}CO_2$ exhalation excretion rates at selected times following a 6-h exposure of male F-344 rats to airborne [^{14}C] formaldehyde at two different concentrations. Each curve was obtained using four rats exposed simultaneously to [^{14}C] formaldehyde vapor. Error bars represent standard deviations.

Table 3   Disposition of Airborne [[14] C] Formaldehyde in F-344 Rats: Percentage Radioactivity in Various Fractions*

Source of radioactivity	Airborne [CH$_2$O], ppm	
	0.63	13.1
Expired air	39.4 ± 1.5	41.9 ± 0.8
Urine	17.6 ± 1.2	17.3 ± 0.6
Feces	4.2 ± 1.5	5.3 ± 1.3
Tissues[†] and carcass	38.9 ± 1.2	35.2 ± 0.5

*Mean ± SD. Four rats (210 ± 4 g) per exposure concentration. Exposure duration was 6 h, and rats were sacrificed 70 h after removal from the exposure chamber.

[†]Nasal mucosa, trachea, esophagus, lung, kidney, liver, intestine, spleen, heart, plasma, erythrocytes, brain, testes.

rapidly over a period of about 12 h. This is followed by a much slower phase, in which the rate of excretion diminishes only slightly with time.

The percentage of the total radioactivity recovered in expired air, urine, and feces, as well as in the tissues and carcasses of the animals sacrificed at 70 h post exposure, is summarized in Table 3. It is evident that the different concentrations of airborne CH$_2$O had little effect on the relative amounts of radioactivity recovered in the different fractions. As expected from other data (1–3), exhalation was the major excretory route of [14] CH$_2$O. However, only about 40 percent of the total radioactivity was eliminated by exhalation. The quantity of radioactivity remaining in the tissues and carcass 70 h after exposure was nearly as large as the total amount excreted by exhalation. Since formaldehyde is a precursor for a large number of biological compounds, the high level of residual radioactivity was probably caused by metabolic incorporation.

Pharmacokinetics

The uptake and disappearance of radioactivity from the blood of rats exposed for 6 h to airborne [14] CH$_2$O were also investigated. A typical result is shown in Fig. 3. Considering first the plasma, it is seen that the concentration of radioactivity increased during the exposure period, reached a maximum at approximately the time of removal from the chamber, and, thereafter, slowly declined over a period of several days. The terminal half-life for radioactivity in the plasma was estimated to be approximately 55 h. Since the half-life for free formaldehyde in rat blood appears to be very short (4, 5), the possibility that this radioactivity is due to formaldehyde can probably be ruled out. The observed half-life of 55 h suggests that [14] CH$_2$O may have been incorporated into the serum proteins, as the half-lives of these proteins in rats are reported to be about 2.9 days (12).

The radioactivity in the packed cell fraction of the blood exhibits a novel,

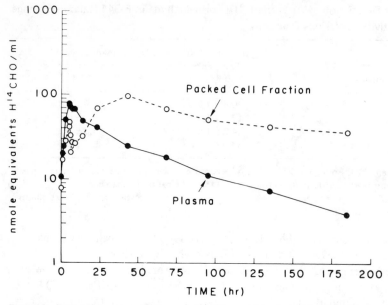

Figure 3 Equivalent concentrations of [^{14}C] formaldehyde in blood (plasma and packed cells) of a male F-344 rat during and after a 6-h exposure to [^{14}C] formaldehyde vapor (8 ppm).

multiphasic kinetic profile (Fig. 3). Following the initial increase in radioactivity that occurs during the exposure, a rapid decline takes place during the first post-exposure hour. Subsequently, the radioactivity in the packed cells again rises, surpasses the plasma concentration, and finally reaches a maximum at approximately 35 h post exposure. In the terminal phase, a very slow decline in packed cell radioactivity takes place, which probably continues for several weeks. The slowness of this decline would be consistent with incorporation of $^{14}CH_2O$ into the erythrocytes, since the half-lives of these cells in rats are reported to be approximately 17 days (13). The increase in radioactivity that precedes the terminal decline is not presently understood. Possible mechanisms include a slow uptake of label by the erythrocytes or, more likely, a rapid, initial labeling of erythrocyte precursors, followed by a slow release of labeled cells into the general circulation.

To understand more clearly the role of metabolism in the incorporation of $^{14}CH_2O$ into blood proteins, rats were injected intravenously either with $^{14}CH_2O$ or with $H^{14}COONa$, and the pharmacokinetics in blood were investigated. Typical kinetic profiles are shown in Fig. 4. These two profiles, one representing formaldehyde injection and the other formate injection, are seen to be quantitatively very similar to each other. After the injection, a rapid decline in radioactivity occurred both in plasma and in the packed cell fraction.

Thereafter, the profiles exhibited the same characteristics as were seen previously, following airborne exposure to $^{14}CH_2O$. The similarities indicate that oxidation of formaldehyde to formate and incorporation of the latter via one-carbon metabolism are of major importance in the long-term pharmacokinetics of $^{14}CH_2O$ following inhalation exposure.

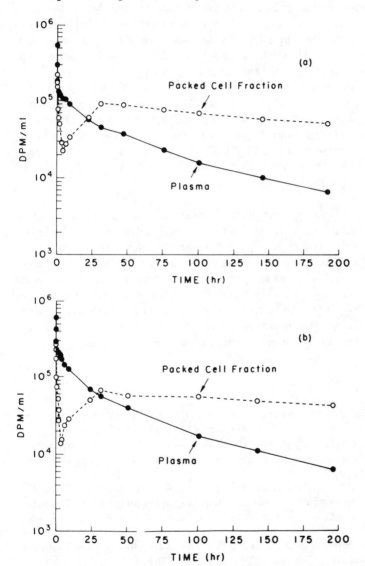

Figure 4 Concentrations of radioactivity in blood (plasma and packed cells) of male F-344 rats following injection of either [^{14}C] formaldehyde (25 μCi) (*a*) or of sodium [^{14}C] formate (29 μCi) (*b*).

DISCUSSION

The results obtained to date demonstrate that airborne $^{14}CH_2O$ is primarily absorbed in the upper respiratory tract, leading to very high local concentrations of radioactivity in the nasal mucosa. Following a 6 h exposure, the amount of $^{14}CH_2O$ absorbed appears to be directly proportional to the airborne concentration. This result is consistent with the high water solubility of CH_2O. It is important as well that the amount of $^{14}CH_2O$ absorbed did not appear to vary following preexposure. Hence, these findings, which are based on single exposures, may also be relevant to the chronic toxicity of CH_2O.

The extensive distribution of radioactivity to other tissues indicates that absorbed $^{14}CH_2O$ or its metabolites are rapidly removed by the mucosal blood supply. It is, however, unlikely that this radioactivity is primarily due to $^{14}CH_2O$. Rapid metabolism in the blood and tissues (6, 8, 14) would probably reduce the concentration of free CH_2O to extremely low levels. In this respect, it is of interest that the testes, a potential target organ for CH_2O adduct formation (15), had among the lowest concentrations of radioactivity of all the tissues examined.

The disposition and pharmacokinetic studies indicate that inhaled CH_2O is extensively metabolized. The high level of residual radioactivity in tissues collected 70 h post exposure and the approximately equivalent amount that is oxidized to $^{14}CO_2$ during the same time period support this concept. It is likely that folic acid plays an important, perhaps a preeminent, role in this incorporation (6). However, the possibility that $^{14}CH_2O$ may form covalent adducts in vivo (15) or that $^{14}CO_2$ may itself be incorporated via carboxylation reactions cannot be excluded. The metabolism of inhaled $^{14}CH_2O$ appears to be similar to that which occurs following other routes of administration (1-3).

REFERENCES

1　Neely, WB. The metabolic fate of formaldehyde-^{14}C intraperitoneally administered to the rat. *Biochem. Pharmacol.* 13:1137–1142 (1964).

2　Buss, J, K Kuschinsky, H Kewitz, and W Koransky. Enterale Resorption von Formaldehyd. *Naunyn-Schmiedebergs Arch. Exp. Pathol. Pharmakol.* 247: 380–381 (1964).

3　Malorny, G, N Rietbrock, and M Schneider. Die Oxydation des Formaldehyds zu Ameisensäure im Blut, ein Beitrag zum Stoffwechsel des Formaldehyds. *Naunyn-Schmiedebergs Arch. Exp. Pathol. Pharmakol.* 250:419–436 (1965).

4　Rietbrock, N. Formaldehydoxydation bei der Ratte. *Naunyn-Schmiedebergs Arch. Exp. Pathol. Pharmakol.* 251:189–190 (1965).

5　Rietbrock, N. Kinetik und Wege des Methanolumsatzes. *Naunyn-Schmiedebergs Arch. Exp. Pathol. Pharmakol.* 263:88–105 (1969).

6   Ntundulu, Th, N Tran, and E Lebel. Détection radiométrique de l'oxydation du formiate [^{14}C] et du formaldéhyde [^{14}C] dans la carence en acide folique. *Arch. Int. Physiol. Biochim.* 84:687–697 (1976).

7   Johansson, EB, and H Tjälve. The distribution of [^{14}C] dimethylnitrosamine in mice. Autoradiographic studies in mice with inhibited and noninhibited dimethylnitrosamine metabolism and a comparison with the distribution of [^{14}C] formaldehyde. *Toxicol. Appl. Pharmacol.* 45:565–575 (1978).

8   McMartin, KE, G Martin-Amat, PE Noker, and TR Tephly. Lack of a role for formaldehyde in methanol poisoning in the monkey. *Biochem. Pharmacol.* 28:645–649 (1979).

9   Swenberg, JA, WD Kerns, RI Mitchell, EJ Gralla, and KL Pavkov. Induction of squamous cell carcinomas of the rat nasal cavity by inhalation exposure to formaldehyde vapor. *Cancer Res.* 40:3398–3402 (1980).

10   Miksch, RR. Formaldehyde in air. A revised NIOSH procedure. Berkeley, CA: Lawrence Berkeley Laboratory, University of California (April 24, 1980).

11   Hebel, R, and MW Stromberg. *Anatomy of the Laboratory Rat*, pp. 55–57. Baltimore: Williams & Wilkins (1976).

12   Esumi, H, S Sato, M Okui, T Sugimura, and S Nagase. Turnover of serum proteins in rats with analbuminemia. *Biochem. Biophys. Res. Commun.* 87: 1191–1199 (1979).

13   Lombardi, MH, and GA Ray. Microtechnique for the study of ferro- and erythrokinetics in the rat. *Am. J. Vet. Res.* 34:253–259 (1973).

14   Uotila, L, and M Koivusalo. Formaldehyde dehydrogenase from human liver. *J. Biol. Chem.* 249:7653–7663 (1974).

15   Stott, WT, and PG Watanabe. Kinetic interaction of chemical mutagens with mouse sperm *in vivo* as it relates to animal mutagenic effects. *Toxicol. Appl. Pharmacol.* 55:411–416 (1980).

Chapter 5

Disposition of [¹⁴C] Formaldehyde after Topical Exposure to Rats, Guinea Pigs, and Monkeys

A. Robert Jeffcoat

Fred Chasalow

Donald B. Feldman

Harry Marr

Human exposure to formaldehyde in industrialized nations is continual and includes dermal contact as well as inhalation and ingestion. Formaldehyde is found in tobacco smoke, in the exhaust of gasoline and diesel engines, and in the biological oxidation products from methylamines. It is used in the manufacture of wood products, as a disinfectant, in the permanent-press process for textiles, and as a copolymer in the production of a number of plastics.

Formaldehyde is an extremely reactive material. It readily adds to most nucleophiles to form hydroxymethylene compounds and reacts with itself to form a variety of low and high molecular weight polymers. It rapidly forms hydroxymethyl compounds with free amino groups in peptides and proteins, which themselves can react further with formaldehyde (1). Formaldehyde reacts with thiols and with a large number of nitrogenous biological building blocks including amino acids, guanidines, purines, and pyrimidines (1-3). In addition to

We wish to thank Mr. Daniel Collins, Ms. Billie Sumrell, and Ms. Susan Mayton for their technical assistance and Drs. John Keller and Joel Bender for their helpful consultation.

This work was supported by a contract with the American Textile Manufacturer's Institute and the Formaldehyde Institute.

the mostly reversible reactions described above, formaldehyde reacts irreversibly with proteins to yield both monomeric and polymeric products (2, 4, 5). Formaldehyde is also readily oxidized to formate and finally to carbon dioxide.

To aid the evaluation of dermal exposure to formaldehyde, the disposition of carbon-14-labeled formaldehyde after topical application to rats, guinea pigs, and monkeys was studied. The results presented in this paper were obtained in the first portion of an ongoing study of the metabolism of formaldehyde following topical administration.

MATERIALS AND METHODS

Adult Fisher 344 rats (Charles River Breeders, Portage, MI) and Dunkin-Hartley guines pigs (Dutchland Farms, Denver, PA) were used. Adult tuberculosis-free cynomolgus monkeys were purchased from Hazleton Primates (Vineland, NJ). After receipt at RTI, the animals were quarantined before use: rats and guinea pigs for at least 7 days, monkeys for 1 month.

The [^{14}C] formaldehyde was purchased from New England Nuclear as an aqueous solution. This material was analyzed by gas chromatography using a Porapak N column at 160°C and by high performance liquid chromatography using a Shodex column with water as the mobile phase. Ninety-six percent of the total radioactivity was eluted in the formaldehyde peak. The [^{14}C] formaldehyde was also diluted with reagent-grade 37% aqueous unlabeled formaldehyde (Fisher Scientific, Raleigh, NC) and the 2,4-dinitrophenylhydrazone derivative formed. From the specific activity of the derivative, the [^{14}C] formaldehyde was calculated to be more than 95 percent pure. The [^{14}C] formaldehyde obtained from New England Nuclear was used directly for the 0.1-mg and 2.0-mg doses and after a 1:3 (v/v) dilution with reagent-grade 37% aqueous formaldehyde (which also contained 10% methanol) for the 11.2-mg dose.

The rats and guinea pigs were kept in individual glass metabolism chambers throughout the experiment. These chambers permitted the collection of urine, feces, and a combination of expired air and evaporation products. On the day prior to the experiment, the portion of the back to which the formaldehyde was to be applied was shaved and a catheter was implanted in the carotid artery of each rat or guinea pig. Each animal was allowed to recover from the surgery overnight. This also permitted any nicks in the shaved portion of the back to heal. An aqueous solution at one of two dose levels of formaldehyde was then applied to a 2-cm^2 area of the shaved portion of the lower back centered along the midline. The lower dose contained 0.1 mg of formaldehyde in 10 μl of solution, while the higher dose contained 11.2 mg of formaldehyde in 40 μl of solution. The amount of carbon-14 in the dose for each experiment was approximately 30 μCi.

Blood samples were collected from each animal at 1, 2, 3, 4, 7, and 24 h after dosing. Urine and feces were collected at daily intervals for 3 days. Airflow through the chambers was maintained at about 600 ml/min. Air was pulled in through the top of the chambers, past the animal, and through an exit at the lower portion of the chamber. Air exiting each chamber was passed through a series of two traps containing 0.5N NaOH. The traps were changed each time a blood sample was taken and at 48 and 72 h after dosing. These traps had previously been shown to trap 100 percent of a sample of $^{14}CO_2$ released in the metabolism chamber and to retain this material for at least 24 h while passing 600 ml/min of air through the traps. At 72 h after dosing, the animals were sacrificed (CO_2) and selected tissues removed. The remaining carcass was then dissolved in NaOH. Blood, excreta, air trap contents, tissue, and carcass were assayed for carbon-14 by scintillation counting either directly or after burning in a Packard Model 306 sample oxidizer. Aliquots of selected air trap contents were each added to a sealed-8 ml vial containing 50% aqueous acetic acid, connected via Teflon tubing to an 8-ml vial containing methanolic 2,4-dinitrophenylhydrazine and finally to a series of two-8 ml vials each containing 5 ml of a specific carbon dioxide absorbant (Carbosorb; Packard Instrument Co.). The amount of $^{14}CO_2$ in the traps was then determined from the amount of carbon-14 contained in the Carbosorb. A single vial containing 5 ml of Carbosorb had previously been shown to trap more than 95 percent of the carbon-14 contained in a solution of $Ba^{14}CO_3$ added to the 50% aqueous acetic acid.

Due to a combination of stress, loss of blood both during surgery and during the blood sampling, and small size (225–300 g), a number of deaths occurred among the F-344 rats during the first day of the experiments. Deaths occurred whether or not the animals had been dosed with formaldehyde. Data collected from animals who died prematurely were not used in the study. Premature death was not a problem with the guinea pigs or with the monkeys.

The studies with monkeys involved several modifications of the above procedure. In each experiment a monkey was placed in a restraining chair. An enclosed cubical Plexiglass hood approximately 40 cm on a side was placed around the monkey's head and fitted snugly around the monkey's neck by means of an overlapping rubber gasket. Air was drawn from the hood, and radiolabeled products were trapped with 0.5N NaOH as described for the experiments using rats and guinea pigs. An aqueous solution containing 2 mg (590–730 μCi) of [^{14}C] formaldehyde in 200 μl of solution was applied to an 18-cm^2 shaved area on the lower portion of the monkey's posterior. The monkey was unable to reach this area while in the restraining chair. Biological samples were collected and analyzed as described for the rats and guinea pigs.

RESULTS

The data obtained when the low dose of formaldehyde (0.1 mg) was applied to guinea pigs are shown in Table 1. Approximately 21 percent of the applied dose was collected in the air traps, most in the first 2h, and about 8 percent was

Table 1 Total ^{14}C Following Dermal Application of 0.1 mg ^{14}CH$_2$O to Guinea Pigs

Time (h)	Air collection data (% dose)								Urine (% dose)			Feces (% dose)			Blood (% dose/total blood volume)					
	0-1	1-2	2-3	3-4	4-7	7-24	24-48	48-72	0-24	24-48	48-72	0-24	24-48	48-72	1	2	3	4	7	24
Males																				
6-10 #1	23.9	2.19	1.24	0.59	0.85	0.93	0.26	0	1.83	0.88	0.36	0.71	0.81	0.74	0.088	0.080	0.072	0.13	0.10	0.096
6-10 #2	14.2	2.16	0.96	S.L.	0.78	0.13	S.L.	0	2.47	0.94	0.46	0.70	0.51	1.06	0.14	0.11	0.15	0.088	0.11	0.14
6-10 #5	18.5	1.45	1.28	0.39	1.02	0.70	0.36	S.L.	1.19	0.87	0.32	0.35	1.65	0.79	0.11	0.12	0.11	0.072	0.052	S.L.
8-26 #1	5.72	5.03	2.44	0.99	1.17	0.60	0.52	0.43	8.66	1.84	2.15	0.92	0.08	0.28	0.15	0.18	0.14	0.13	0.11	0.13
8-12 #2	18.4	3.24	1.76	0.94	1.40	1.77	0.70	0.41	3.08	3.28	1.40	0.41	0.13	N.S.	0.078	0.099	0.073	0.088	0.073	0.083
Females																				
6-17 #1	9.24	3.08	0.91	0.98	1.95	1.40	0.50	0.32	1.18	0.31	0.22	0.15	0.80	0.56	0.046	0.072	0.061	0.027	0.038	0.046
6-17 #2	17.4	1.67	0.91	0.71	0.92	1.18	0.54	0.28	0.77	0.58	0.97	N.S.	0.35	0.33	0.25	0.26	0.30	0.023	S.L.	0.14
7-28 #2	8.61	1.69	1.07	0.66	1.13	1.70	0.92	0.39	0.79	0.59	1.65	0.29	0.24	0.57	0.024	0.040	0.036	0.032	0.052	0.064
7-28 #4	12.4	1.21	0.70	0.46	1.17	0.92	0.39	0.32	1.31	1.74	2.29	0.21	0.11	0.49	0.032	0.018	S.L.	0.040	0.054	0.068
8-12 #1	15.3	3.99	2.38	1.41	2.37	2.88	0.82	0.47	0.96	0.43	0.48	0.23	0.35	0.24	0.088	0.072	0.088	0.080	0.056	0.072
8-12 #3	7.37	3.20	0.89	0.45	1.34	1.40	0.84	0.62	2.07	2.14	1.65	0.59	0.45	0.17	0.16	0.18	0.16	0.16	0.21	0.18

S.L. = sample lost; N.S. = no sample.

excreted in the urine and feces. Further examination of the material in the air traps showed that only a small proportion (<3 percent) was present as CO_2. Analysis of the trapped material for formaldehyde was not possible due to the instability for formaldehyde in the NaOH collection medium. The concentration of formaldehyde-derived material in the blood was fairly constant, averaging 0.10 percent of the applied dose in the total estimated blood volume of the animal throughout the data collection period. No significant differences between sexes were observed in any of the data. Table 2 shows the distribution of radiolebeled material at 72 h after application of the formaldehyde. No large accumulation of radiolabel occurred in any internal organ, but the site of application retained 16 percent of the original carbon-14. The carcass remaining after removal of the internal organs contained 27 percent. A similar pattern was observed with rats administered the same amount of formaldehyde (Table 3).

Table 4 shows the results when the high dose of formaldehyde (11.2 mg) was applied to guinea pigs. The percentage of carbon-14 remaining at the site of application after the higher dose (about 4 percent) was much lower than the 16 percent remaining after application of the lower dose, although the mass of labeled material remaining was actually higher. The data collected from rats (Table 5) are again quite similar. Several hours after application of the 11.2-mg dose of formaldehyde to rats, a yellow color developed in the fur surrounding the site of application. The yellow color then gradually diffused radially outward into the fur over the next 24 h. This yellow coloring was only slightly apparent on the guinea pigs after application of the 11.2-mg dose of formaldehyde and was not visible at all after application of the lower dose to either species.

The results obtained from the studies with monkeys (Table 6) are somewhat different from those obtained from the rodents. These animals, as might be expected due to the much greater heterogeneity of their breeding as compared to the rodents, also showed a much higher animal to animal variation. In all instances, however, the sum of excreta (air, urine, and feces) from the monkeys was less than 1 percent of the applied dose. Carbon-14 remaining in the internal organs of the monkeys 72 h after dosing was quite low and fairly evenly distributed. The carbon-14 found in all of the organs examined totaled less than 0.05 percent of the applied dose.

For comparison, the averaged results from all three species are presented in Tables 7 and 8. Other than the decreased percentage of the radiolevel remaining at the site of application in the animals dosed with 11.2 mg of formaldehyde, there are no significant differences in the disposition of the radiolevel between doses or between the two rodent species. Because of the different collection system, the air traps from the monkeys contained only carbon-14 from expired air. The percentage of the applied carbon-14 contained in the air traps from the monkeys is less than 2 percent of the total radioactivity or about 10 percent of the $^{14}CO_2$ collected from the rodents. The percentage of the dose excreted in

Table 2 Tissue Distribution of ^{14}C 72 h after Topical Administration of 0.1 mg ^{14}CH$_2$O to Guinea Pigs[*]

	Heart	Lung	Liver	Spleen	Kidney	Leg (per g)	Brain	Gonad	Skin (site of application)	Skin & hair far removed from site of application (per g)	Remaining carcass
Males											
6-10 #1	0.014	0.071	0.194	0.014	0.055	0.007	0.009	0.010	2.76	0.032	30.0
6-10 #2	0.049	0.120	0.136	0.014	0.092	0.002	0.012	0.011	1.52	0.044	25.6
6-10 #5	0.015	0.042	0.124	0.031	0.046	0.009	0.008	0.007	23.0	0.223	16.6
8-26 #1									16.5	0.261	27.9
8-12 #2									17.9	0.263	36.3
Females											
6-17 #1	0.012	0.038	0.114	0.004	0.034	0.003	0.006	0.001	7.7	0.057	25.6
6-17 #2	0.044	0.128	0.411	0.066	0.143	0.014	0.031	0.004	14.6	0.092	28.6
7-28 #2	0.011	0.032	0.077	0.009	0.031	0.003	0.005	0.002	16.7	0.219	22.7
7-28 #4	0.010	0.047	0.114	0.013	0.036	0.003	0.004	0.003	20.6	0.174	21.4
8-12 #1									22.1	0.242	29.1
8-12 #3									27.9	0.198	34.9

[*]Data presented as percent of applied dose.

Table 3   Distribution of ^{14}C during the First 73 h after Topical Administration of 0.1 mg ^{14}CH$_2$O to Rats[*]

	Air traps	Urine	Feces	Skin (site of application)	Remaining carcass	Mean blood content[†]
Males						
7–16 #4	30.7	2.4	0.1	14.2	20.9	0.13
8–26 #6	32.8	6.1	3.4	15.6	17.2	0.061
8-12 #4	30.7	7.8	4.2	16.1	26.5	0.15
8–12 #5	39.9	3.9	0.1	23.1	20.4	0.11
Females						
6–24 #2	21.7	2.9	0.9	15.8	26.4	0.13
6–24 #3	21.6	6.5	1.6	10.4	22.5	0.18
8–26 #8	35.6	5.4	0.2	21.9	17.2	0.081
9–26 #4	18.8	4.2	2.7	11.8	24.3	0.14
9–26 #6	23.3	6.0	1.1	17.3	24.7	0.12

[*]Data presented as % of applied dose.
[†]Average % dose in total blood volume over 1–24 h.

Table 4   Distribution of ^{14}C during the First 72 h after Topical Administration of 11.2 mg ^{14}CH$_2$O to Guinea Pigs[*]

	Air traps	Urine	Feces	Skin (site of application)	Remaining carcass	Mean blood content[†]
Males						
7–22 #1	17.5	15.4	1.0	3.72	24.2	0.062
7–22 #2	20.1	9.2	1.4	2.92	22.9	0.057
7–22 #4	26.8	5.4	1.0	4.10	33.6	0.051
7–22 #5	47.5	2.8	0.7	6.45	19.2	0.049
8–22 #1	9.3	7.2	1.1	1.66	27.3	0.096
8–22 #6	13.6	2.8	2.1	4.06	26.9	0.14
Females						
8–1 #3	23.3	5.5	1.2	5.50	29.4	0.14
8–1 #4	21.2	8.8	0.6	4.80	31.4	0.10
8–1 #5	32.9	9.4	0.8	0.66	26.4	0.052
9–26 #1	29.9	3.1	2.1	3.89	33.4	0.094
9–26 #2	19.4	5.1	0.8	4.05	37.2	0.12

[*]Data presented at % of applied dose.
[†]Average % dose in total blood volume over 1–24 h.

Table 5 Distribution of ^{14}C during the First 72 h after Topical Administration of 11.2 mg ^{14}CH$_2$O to Rats[*]

	Air traps	Urine	Feces	Skin (site of application)	Remaining carcass	Mean blood content[†]
Males						
8-1 #1	26.3	4.1	0.6	2.8	33.2	0.11
8-22 #5	12.8	8.7	0.6	2.5	24.2	0.14
9-26 #7	23.5	11.8	1.0	4.0	25.0	0.16
Females						
8-18 #4	18.4	9.8	—	4.5	31.0	0.11
8-18 #7	33.3	4.1	0.8	5.3	17.4	0.12
8-18 #8	21.9	9.1	0.3	3.2	31.4	0.16
8-22 #8	12.2	11.0	1.3	1.6	23.3	0.077
9-26 #3	28.1	7.7	1.0	3.6	21.3	0.14

[*]Data presented as % of applied dose.
[†]Average % dose in total blood volume over 1–24 h.

the urine and feces of the monkeys was also approximately 10 percent of that excreted by the rodents. The percentage of the dose remaining at the site of application after applying 2 mg of formaldehyde to monkeys was intermediate between that remaining at the site of application in the rodents that received 0.1 mg and those who received 11.2 mg. The fractional dose in the blood circulation (Table 8) in the monkey was also substantially less than that observed with the rodents. The carbon-14 content in the blood determined for several rats 72 h after application of [^{14}C] formaldehyde was approximately the same as that determined during the first 24 h after dosing.

At the time of sacrifice, 72 h after drug administration, very little carbon-14 was found in the major organs excised during necropsy. The internal organs containing the highest concentrations of carbon-14 were from animals that had been given 11.2 mg formaldehyde. The total in all of these internal organs was less than 1 percent of the applied dose. Concentrations of carbon-14 in the internal

Table 6 Tissue Distribution of ^{14}C during First 72 h after Topical Administration of 2.0 mg ^{14}CH$_2$O to Monkeys[*]

	Air traps	Urine	Feces	Skin (site of application)	Mean blood content[†]
#1	0.64	0.18	0.02	17.0	0.024
#2	0.06	0.44	0.42	3.91	0.0046
#3	0.42	0.09	0.16	7.55	0.017

[*]Data presented as % of applied dose.
[†]Average % dose in total blood volume over 1–24 h.

Table 7 Mean Values for the Distribution of ^{14}C during the First 72 h after Topical Administration[*]

Species	Dose (mg)	Air traps	Urine	Feces	Skin (site of application)	Remaining carcass	Total ^{14}C recovered	Mean blood content[†]
Rat	0.10	28.3 ± 2.4	5.0 ± 0.6	1.5 ± 0.5	16.2 ± 1.4	22.2 ± 1.2	73.4 ± 3.1	0.12 ± 0.01
Guinea pig	0.10	21.4 ± 1.6	4.5 ± 1.0	1.4 ± 0.2	15.6 ± 2.5	27.1 ± 1.7	70.0 ± 3.7	0.10 ± 0.02
Rat	11.2	22.1 ± 2.6	8.3 ± 1.0	0.7 ± 0.1	3.4 ± 0.4	25.9 ± 1.9	60.4 ± 2.6	0.13 ± 0.01
Guinea pig	11.2	23.8 ± 3.1	6.8 ± 1.1	1.2 ± 0.4	3.8 ± 0.5	28.4 ± 1.6	63.6 ± 2.6	0.09 ± 0.01
Monkey	2.0	0.37 ± 0.17	0.24 ± 0.10	0.20 ± 0.12	9.49 ± 3.90	[‡]		0.015 ± 0.0006

[*]Data presented as % of applied dose ± S.E.
[†]Average of % dose in total blood volume over 1–24 h.
[‡]Not analyzed.

Table 8 Mean Values for ^{14}C in the Total Estimated Blood Volume as a Function of Time[*]

Species	Dose (mg)	1 h	2 h	3 h	4 h	7 h	24 h
Rat	0.10	0.07 ± 0.01	0.10 ± 0.01	0.13 ± 0.01	0.13 ± 0.01	0.13 ± 0.01	0.13 ± 0.02
Guinea pig	0.10	0.11 ± 0.02	0.11 ± 0.02	0.12 ± 0.02	0.08 ± 0.10	0.09 ± 0.02	0.10 ± 0.01
Rat	11.2	0.08 ± 0.01	0.10 ± 0.01	0.12 ± 0.01	0.14 ± 0.01	0.15 ± 0.02	0.17 ± 0.03
Guinea pig	11.2	0.17 ± 0.05	0.10 ± 0.01	0.09 ± 0.01	0.09 ± 0.01	0.07 ± 0.01	0.09 ± 0.02
Monkey	2.0	0.006[†]	0.015 ± 0.005	0.013 ± 0.005	0.015 ± 0.006	0.018 ± 0.008	0.021 ± 0.011

[*]Data presented as % dose ± S.E.
[†]Average of two values; range was ±0.002.

organs of the animals given about 1 percent of this dose were correspondingly smaller. In contrast, significant carbon-14 was found at the site of application, on the skin and hair sample taken from the chest of the animal (average 0.41 percent of the dose per gram for the rats; 0.16 percent for the guinea pigs; and 0.004 percent for the monkeys), and in the remainder of the carcass.

In one rat, additional hair and skin samples were assayed. Hair clipped from skin 1–2 cm from the site of application contained 1.2 percent of the dose of carbon-14, while the skin plus the hair stubble 0–2 cm from the application site contained 0.9 percent of the dose. Hair removed from the major portion of the remainder of the rat contained 5.3 percent of the dose. Another 5.5 percent of the dose was contained in the major portion of the remaining skin. After removal of this hair, skin, and the internal organs, 11 percent of the applied carbon-14 was found in the remaining carcass.

DISCUSSION

In our studies, an average of 6.6 percent of the dose applied to rodents was excreted in the urine over 72 h, while less than 3 percent was expired as CO_2. For the monkeys, approximately equal amounts were excreted by the two routes, both less than 1 percent of the applied dose. These results are drastically different from those found after administration of radiolabeled formaldehyde internally. Buss et al. (6) found about 40 percent of the dose as expired $^{14}CO_2$ within 12 h after oral administration of $[^{14}C]$formaldehyde to rats. During this period, an additional 10 percent was excreted in the urine. In another study in which the $[^{14}C]$formaldehyde was given intraperitoneally to rats, 82 percent of the dose was expired as $^{14}CO_2$ within 48 h and 13 percent was excreted in the urine in the form of methionine, serine, and an adduct with cysteine (7). Rapid conversion of intravenously administered $[^{14}C]$formaldehyde to formate has been found in dogs, with somewhat slower metabolism of the formate ($t_{1/2} = 82$ min) (8). After intravenous administration of $[^{14}C]$formaldehyde to mice, an average of 46 percent of the radiolabel was exhaled as $^{14}CO_2$ within 1 h (9). In all studies, the majority of the radiolabel was metabolized by oxidation to $[^{14}C]$formate and finally to $^{14}CO_2$. When $[^{14}C]$formaldehyde was given orally and intraperitoneally to rats, four to six times as much carbon-14 was expired as $^{14}CO_2$ as was excreted in the urine over the same time period. This indicates that there are fundamental differences in the disposition of formaldehyde topically applied and that taken directly into the body. We believe that the major differences in the disposition occur because formaldehyde is free to evaporate from the skin either directly or after disassociation from reversible complexes. Thus, a large portion of the carbon-14 that enters the systemic circulation after topical application would be products of either the irreversible bonding of formaldehyde to biomolecules or incorporation of the radiolabel via the one-carbon pool into compounds such as methionine and serine. These compounds are eliminated by urine or feces rather than by the breath.

The interaction of topically applied formaldehyde with an animal can be viewed as follows. The formaldehyde can either a) evaporate from the skin surface, b) pass through the skin, or c) react with biomolecules in the skin reversibly or irreversibly. Reversible reaction with skin components leads to a slow release of formaldehyde from the reaction site or its later release after the formaldehyde-biomolecule complex has moved away from the initial site of reaction. Finally, formaldehyde can be metabolized by oxidation to formate and then to CO_2. In our studies with rats and guinea pigs, an average of 21–28 percent of the applied radiolabel was recovered from the air traps (Table 7), the majority within 2 h after dosing. It appears, however, that the vast majority of the material contained in the air traps came from evaporation of formaldehyde from the skin surface. This interpretation is consistent with the results from our studies with the monkey. [^{14}C] formaldehyde that evaporated from the monkey skin would not be collected with the expired air, and, as noted, the amount of labeled material collected in this way was about 60 times smaller than for the rodents.

Our data indicate, therefore, that after application to the skin, most of the formaldehyde evaporates. Before the formaldehyde vapors diffuse very far from the site of application, a substantial portion reacts with biomolecules in the nearby hair and skin, in both reversible and irreversible processes. As time goes on, formaldehyde generated from the reversible reactions migrates farther away from the site of application. When a large enough amount of formaldehyde is present, this migration can be seen as a moving yellow band. This explains the high concentration of carbon-14 in the hair and skin of the rat and the guinea pig. Since the monkey is less hairy and is sitting in a chair, which shields a sizable portion of his body, a much smaller portion of the carbon-14 would be expected in the skin and hair.

The carbon-14 present in the carcass of the rat remaining after removal of the major internal organs and most of the skin and hair may be due to diffusion of radiolabeled material into the tissues below the site of application. The exact location of this material is currently being studied.

The evaporation of formaldehyde from the site of application could possibly lead to the rat and guinea pig being exposed to very limited amounts of formaldehyde by inhalation. Calculations based on the amount of carbon-14 contained in the air traps and on the respiration rates and tidal volumes of these animals limit inhalation to a few percent at most of the administered dose. Not considered in these calculations is the necessity for the formaldehyde to diffuse against the flow of air through the metabolism chamber.

The total recovery of carbon-14 in the studies with rats and guinea pigs averaged approximately 67 percent. The remaining radiolabel probably evaporated between the time that the dose was applied to the animal and the animal was placed in the metabolism chamber. This time interval was approximately 15 s. The recovery of carbon-14 from the monkey was only 10 percent. This reflects the fact that none of the evaporated formaldehyde was recovered, that the carbon-14 in the remaining carcass was not measured, and that the entry of the carbon-14 into the systemic circulation was much lower than with the rodents.

In summary, we have shown that less than 1 percent of the applied dose of carbon-14 is excreted or concentrated in the major organs of the monkey when [^{14}C] formaldehyde is applied topically. Approximately ten times this amount was found in rat and guinea pig excreta and internal organs. Coupled with the much lower blood levels of carbon-14 found in the monkey, these results indicate that the skin of the monkey is much less permeable to formaldehyde than that of the rodents. Significant proportions of the dose are found after 72 h at the site of application, in the skin and fur, and, at least for the rodents, in the remaining carcass. The large majority of the applied radiolabel in all cases is lost to evaporation.

In the second phase of our ongoing study, the interaction of [^{14}C] formaldehyde with skin, blood, and liver will be studied in more detail. The rates of metabolism of the formaldehyde and the nature of the metabolic products will be ascertained in in vitro experiments using tissues from rats, monkeys, and humans. In the final phase of the study, monkeys will be fitted with a garmet impregnated with a resin containing [^{14}C] formaldehyde. The disposition of carbon-14 absorbed from the fabric will be determined with the same techniques as used in the first phase of the study.

REFERENCES

1 Kallen, RG, and VP Jencks. Equilibria for the reaction of amines with formaldehyde and protons in aqueous solution. *J. Biol. Chem.* 241:5864–5878 (1966).

2 Galembeck, F, DS Ryan, JR Whitaker, and RE Feeney. Reactions of proteins with formaldehyde in the presence and absence of sodium borohydride. *J. Agri. Food. Chem.* 25:238–245 (1977).

3 Feeney, RE, G Blankenhorn, and HBF Dixon. Reactions of carbonyl compounds with amino acids and proteins. In *Advances in Protein Chemistry*, edited by CB Anfinsen, JT Edsall, FM Richards, vol. 29, pp. 149–150. New York: Academic Press (1975).

4 Myers, JS, and JK Hardman. Formaldehyde-induced cross-linkages in the α-subunit of the *Escherichia coli* tryptophan synthetase. *J. Biol. Chem.* 246: 2863–3869 (1971).

5 Warren, JR, L Spero, and JF Metzger. The pH dependence of enterotoxin polymerization by formaldehyde. *Biochem. Biophys. Acta* 365:434–428 (1974).

6 Buss, J, HK Kuschinsky, and W Koransky. Enterale resorption von formaldehyd. *Arch. Exp. Pathol. Pharmacol.* 247:380–381 (1964).

7 Neely, WB. The metabolic fate of formaldehyde-^{14}C intraperitoneally administered to the rat. *Biochem. Pharmacol.* 13:1137–1142 (1964).

8 Malorny, G, N. Rietbrock, and M Schneider. Die oxydation des formaldehyds zu ameisensaüre im blut, ein beitrag zum stoffwechsel des formaldehyds. *Arch. Exp. Pathol. Pharmacol.* 250:419–436 (1965).
9 Johansson, EB, and H Tjälve. The distribution of [^{14}C] dimethylnitrosamine
   in mice. Autoradiographic studies in mice with inhibited and noninhibited
   dimethylnitrosamine metabolism and a comparison with the distribution of
   [^{14}C] formaldehyde. *Toxicol. Appl. Pharmacol.* 45:565–575 (1978).

Chapter 6

Teratogenicity of Formaldehyde

R. E. Staples

FORMALDEHYDE STUDIES

Connors (1) listed formaldehyde as a teratogen and gave site of action as co-valent reaction with DNA, inhibition of DNA synthesis, or inhibition of DNA repair. This was based on the reduced DNA content measured in the rat fetus by Pushkina et al. (2). In this study, rats were exposed to formaldehyde vapor for 24 h/day from 20 days pre-mating, during cohabitation, and throughout pregnancy at exposure levels of 0, 0.012, and 1.0 mg/m^3. At term the dams were decapitated. The fetuses, placentas, and maternal and fetal livers were analyzed for ascorbic acid and nucleic acid levels according to methods usually used in the U.S.S.R. in 1963. No external malformations were visible among the fetuses. In the groups exposed to CH$_2$O the number of fetuses per litter was reduced and fetal weight was reported as being increased, but these differences were apparently not statistically significant. The article was uninformative regarding maternal clinical signs, body weight gain, resorptions, and visceral or skeletal alterations, and no data were presented regarding the nucleic acid levels in either the dams or the fetuses, but the results obtained were interpreted by Pushkina et al. (2) as constituting an embryotropic effect.

A teratogenic response consists of an alteration that occurs during the course of development to result in a permanent structural or functional defect. The abnormality referred to by Connors (1) was a reduced DNA content for whole fetuses from formaldehyde-exposed dams. The permanency of this change and the alteration in ascorbic acid levels claimed by Pushkina et al. (2) were not validated by following continued development of any of the offspring. Hence, as presented, the data do not support the presence of a teratogenic response but represent a toxic response at best. In this regard, it is of interest to note that in a study by Ritter et al. (3) a 25–75 percent suppression of DNA synthesis in the rat conceptus for up to 9 h on day 12 of gestation left little effect apparent in fetuses on day 20 of gestation.

Several additional studies were conducted on the teratogenicity or embryotoxicity of formaldehyde. Gofmekler (4) also exposed rats to formaldehyde at levels of 0, 0.012, and 1.0 mg/cubic meter for 24 h/day by inhalation but for only 10–15 days before mating, as well as during the mating period, and throughout gestation. In the 1.0 mg/cubic meter group, fetal weight and fetal adrenal, kidney, and thymus weights increased, but fetal lung and liver weights decreased. In the 0.012 mg/cubic meter group only the increase in fetal weight and adrenal weight was noted. Although not mentioned by the author, it appears as though formaldehyde exposure increased the number of pups per litter. Statistical analyses were not applied to any of the data presented so it is not known whether the reported weight changes represented significant differences. Similarly, the day of mating was not identified for any female during the 10-day mating period; hence, the claimed 14–15 percent increase in duration of gestation was not demonstrated.

Pathological examination of these fetuses (5) revealed no external malformations, no macroscopic structural changes of the internal organs or the skeletons, and no developmental delay. The histologic changes noted in the 36 fetuses of the 1.0 mg/cubic meter group that were studied further included larger extramedullary hematopoietic centers, epithelial proliferation in the common bile ducts, polymorphism of renal epithelial cells, cast-off cells in the lumina of some tubules, decreased myocardial glycogen, involution of thymic lymphoid tissue, and disintegration of lymphocytes. These are not unusual lesions but represent the expected types of changes that should also have been present in the control fetuses. No quantitative data were given; therefore, all of the changes presented were evaluated subjectively. The study apparently was not coded, and there was no evidence of technical controls being conducted to minimize bias in demonstration of effects across groups. Hence, the effects claimed were not convincingly demonstrated.

Guseva (6) exposed male rats to formaldehyde by two routes simultaneously: by inclusion in drinking water for 5 days per week and by inhalation for 4 h/day for 5 days per week over a period of 6 months with three dose levels per route. At the two highest regimens, i.e., at 0.1 mg/liter in drinking water with 0.5 mg/cubic meter by inhalation and at 0.01 mg/liter and 0.25

mg/cubic meter, respectively, a significant decrease in testicular nucleic acid level was measured in the male rats. The lowest dose level tested, i.e., 0.005 mg/liter and 0.12 mg/cubic meter, did not adversely affect testicular nucleic acid level. If the variability between the data presented for the control level and the data for the lowest exposure level for which no adverse effect was reported is indicative of the sensitivity of the method or of the degree of uniformity among duplicate groups, then the validity of the reported decrease in the higher dose groups may be questioned.

The reproductive function of the rats in the high and low exposure levels was assessed by breeding them to nonexposed female rats. However, each group contained only three male rats (12 in the study), each of which was bred to two females (six females per group), and then some of these females in each group were sacrificed at term to determine their reproductive status. The remainder gave birth and reared their offspring for one month during which external landmarks were noted to assess rate of development. No adverse effects were noted in fetal body weight, in the incidence of birth defects, or in developmental anomalies. All of the females mated to males in the low dose group became pregnant, but the article was uninformative regarding the pregnancy rate for the high dose group. In any case, the number of animals per smallest subgroup was not sufficient to convincingly demonstrate adverse formaldehyde-related effects on the criteria studied.

Kalmykova [7] reported symmetric translocations at diakinesis of metaphase I in the form of rings and chains in the testes of mice at both 24 h and 12 days post exposure to milk containing 0.3 ml Formalin per liter—a dose of 17 mg/kg (1/25 of the LD_{50}) for 2 months. This finding may represent a significant mutagenic event and needs to be evaluated, but I would like to address the reported "significant increase in postimplantation mortality" in the offspring of unexposed dams bred to the above males after the dosing period. The increase in postimplantation mortality at the secondary spermatocyte stage appeared to be due to an unusually high implantation rate rather than to an adverse effect of Formalin exposure as reported by the author. The 80 implants for this group of ten females was by far higher than that for the control group and for all of the remaining groups, and the percentage of live embryos was not reduced below the control value.

Ranstroem [8] subcutaneously injected 0.25 ml X 2 of a 6% Formalin solution into rats daily throughout gestation. He reported an increase in fetal weight (5.1 g vs 4.7 g) and a decrease in adrenal weight, particularly in the small fetuses. The author interpreted the changes as being due to a change in endocrine function. No additional data were given, e.g., number of females per group, number of fetuses per litter, variability in fetal weight in the experimental group versus the control group. The reader, therefore, cannot independently evaluate the biological significance of the finding even if it was repeatable.

Isaacson and Chaudhry [9] injected strain A mice twice, 6 h apart on day

11.5 of gestation, with 0.15 ml of a 2% formaldehyde solution intramuscularly and observed 1 of 88 live fetuses to have cleft palate alone. Two additional fetuses with cleft palate alone were observed among the litters of six females in a previous study after a single injection of 0.3 ml of a 4% formaldehyde solution. The authors recommended further study on the basis of cleft palate alone being a rare finding in this strain of mice. An incidence of 0.3 percent was quoted from a study conducted elsewhere. I found no evidence of this study having been repeated in the interval since 1962. The results of this paper as presented certainly do not demonstrate formaldehyde to be a teratogen in the rat.

Hurni and Ohden (10) added formaldehyde (40% solution) to the diet of Beagle dogs at 125 and 375 ppm (3.1 and 9.4 mg/kg body weight/day) from days 4 through 56 of gestation. No adverse effects were noted among the nine or ten pregnant dams per group or among their offspring after birth. Criteria evaluated included pregnancy rate, maternal weight gain, length of gestation, number of fetuses per litter, and external, visceral, or skeletal malformations in the stillborn pups and in those that died before weaning.

Marks et al. (11) administered formaldehyde diluted in distilled water to CD-1 mice by gavage from days 6 through 15 of gestation at dose levels that ranged from one that was lethal to more than one-half of the dams to one that had no apparent effect on the concepti. The dams were coded before sacrifice so the laboratory personnel did not know which compound was being tested or which were the experimental or control groups. There were 76 mated females in the control group that were gavaged with distilled water and at least 29 mated females were in each of the remaining groups. Twenty-two of the thirty-four dams in the high dose group (185 mg formaldehyde/kg/day) died before scheduled sacrifice. The deaths occurred between days 8 and 15 of gestation, usually after 1 or 2 days of extensive body weight loss. As a result, only seven litters were available for study in the high dose group. However, no compound-related difference was noted in the maternal body weight gain between days 6 and 17 of gestation for the dams of all groups that survived to scheduled sacrifice on day 18 of gestation. The same was true also for incidence of implants, sex ratio, stunted fetuses, average number of live fetuses per litter, and average fetal weight per litter. An increase in the incidence of resorptions was noted in the high dose group, but the difference from the control value was not statistically significant. All of the fetuses were examined for external and skeletal alterations, and one-third of the fetuses of each litter were examined for visceral alterations. Therefore, even at dose levels lethal to a large proportion of the dams (185 mg/kg/day), formaldehyde was not demonstrated to be teratogenic in the CD-1 mouse, and embryotoxicity was not noted at the 145 mg/kg/day dose level or below.

Methanol was present in the stock formaldehyde solution at 12–15% as a stabilizer, but methanol was not added to the distilled water given to the control dams by gavage. However, the amount of methanol present was miniscule in comparison to the dose required to demonstrate toxicity in the adult rat (12, 13). Hence, it is highly unlikely that the maternal lethality observed was due to the methanol rather than to the formaldehyde received.

HEXAMETHYLENETETRAMINE STUDIES

As it apparently has not been demonstrated that formaldehyde or its toxic metabolites ever reach the conceptus after administration by the oral, dermal, or inhalation route, I became interested in the results of teratogenicity studies conducted on two possible donors of formaldehyde in vivo. The first, hexamethylenetetramine, was fed to Beagle dogs by Hurni and Ohder (10) at 600 and 1250 ppm from days 4 through 56 of pregnancy. No malformed pups were noted, but of the 56 pups born to dams in the high dose group, ten were stillborn. Seven of these occurred in one litter. Between birth and weaning the growth rate of the surviving pups was reduced, and a significant increase in mortality occurred in comparison to the control value. Hence, although a teratogenic response was not demonstrated, a toxic response was noted at the high dose level of 1250 ppm in the diet.

In another study, Della Porta et al. (14) gave six male and 12 female Wistar rats 1% hexamethylenetetramine in their drinking water for 2 weeks before mating and to the mated females through pregnancy and lactation. External malformations were not noted among the 124 pups born to these dams or among the 118 from the 11 untreated control dams. A temporary decrease in body weight gain was noted on continued exposure of the pups to formaldehyde for 20 weeks post partum. Since the decrease was temporary, this represented an embryotoxic rather than a teratogenic response. Although a teratogenic response was not demonstrated, it is important to note that only one dose level was given and that malformed pups might have been cannibalized if present. However, the number of live pups born per litter was comparable in the formaldehyde and control groups.

In another reproduction study, Natvig et al. (15) incorporated hexamethylenetetramine into the diet of 2-month-old rats of both sexes at a concentration of 0.16 percent (1600 ppm) with the primary purpose of testing its toxicity as a food additive. The exposed group consumed about 100 mg/kg/day. After 3 months of exposure, the 16 males and 16 females that formed the experimental group were interbred. No differences were detected between the hexamethylenetetramine-exposed group and in the control group in weight gain, running activity, and the fertility of the first or second generation.

Therefore, the only possible adverse effects found to date consist of some embryotoxicity at 31 mg/kg/day by diet to dogs and in the rat after prolonged consumption of 2 g/kg/day in the drinking water.

HEXAMETHYLPHOSPHORAMIDE STUDIES

A second potential formaldehyde donor, hexamethylphosphoramide is meta-bolized more slowly in vivo than is hexamethylenetetramine since a large propor-tion of an ingested labeled dose appeared in the urine as unchanged hexa-methylphosphoramide (16). It is possible then that formaldehyde may be released after hexamethylphosphoramide is distributed into the environment of the conceptus. Therefore, teratogenicity studies conducted on this chemical could be pertinent to the present subject even though it is recognized that the etiology of any adverse findings could be due to formaldehyde or to the hexa-methylphosphoramide per se. Hexamethylphosphoramide is a known chemo-sterilant, but a teratogenic response was not observed in two oral studies with rats.

Kimbrough and Gains (17) administered hexamethylphosphoramide to ten female rats by gavage at the dose level of 200 mg/kg/day for 7 days before mating them to unexposed male rats. The females then continued to be gavaged with hexamethylphosphoramide until sacrifice on day 20 of gestation. The fetuses were examined for external and visceral structural alterations. No mal-formations were noted in fetuses from the group given hexamethylphosphora-mide or among the fetuses of eight control females gavaged with water only. No significant differences were noted between groups in fetal or placental weight, number of fetuses per litter, or in the incidence of resorptions.

Shott et al. (18) administered hexamethylphosphoramide to rats by gavage at 2 or 10 mg/kg/day from the beginning of the breeding period to weaning of the F_{1A} offspring. No adverse effects were noted among the F_{1A} offspring or among the nonexposed offspring. Similarly, after atmospheric exposure to 0.344 ppm of hexamethylphosphoramide for 6 hours per day from days 6 through 15 of gestation, neither embryotoxicity nor teratogenicity was noted (Haskell Laboratory Report, unpublished data).

Therefore, hexamethylphosphoramide appeared to not adversely affect rat concepti when given by gavage at up to 200 mg/kg/day or by inhalation at 0.344 ppm.

GLYCEROL FORMAL STUDIES

A condensation product of glycerin and formaldehyde, glycerol formal, a com-monly used solvent for toxicity testing at least in Italy, was reported to be teratogenic after administration subcutaneously or intramuscularly to the rat at all dose levels from days 6 through 15 of gestation (19, 20). At sacrifice on day 21 of gestation, up to 76 of 193 fetuses in the experimental group were malformed, and a positive dose response was demonstrated. The malformations consisted mostly of ventricular septal defects, aortic arch malformations, and wavy ribs. Additional study demonstrated that the heart and great vessel malformations resulted from exposure to glycerol formal between days 10 and 12 of gestation, whereas the rib malformations occurred after any 2- or 3-day exposure period between days 7 and 16 of gestation (21). It is not known whether humans are likely to be exposed to this condensation product or whether the product can be formed in vivo; therefore, its significance as a hazard to the human conceptus is unknown.

CONCLUSIONS

It is important to first point out that the significance of the results for several of the studies discussed are difficult if not impossible to evaluate not only because the number of animals included in both experimental and control groups often do not meet minimal requirements but also because the data and associated information presented are not sufficiently complete to permit independent evaluation. In addition, the integrity of the test chemical and the sensitivity and accuracy of chemical and biochemical analyses often are not documented. The results claimed in articles with such deficiencies should be given little if any weight during collection and review of data for assessing degree of hazard associated with exposure.

Second, formaldehyde has not been demonstrated to be teratogenic in any species to date, but data of value for estimating risk to the human conceptus is quite limited. Useful published data probably do not extend beyond the oral route of administration in the dog and mouse.

Third, a study demonstrating that formaldehyde or formic acid per se reaches the mammalian conceptus by the oral, dermal, or inhalation route has not been identified to date.

Fourth, glycerol formal has been demonstrated to be a potent teratogen in the rat at dose levels that were not toxic to the dam.

RECOMMENDATIONS

1 That a teratogenicity study by inhalation be conducted on formaldehyde, formic acid, or both in a pertinent test species. The protocol should relate to current scientific knowledge and be in compliance with the spirit of the proposed TSCA guidelines.

2 That if a chemical can be identified which will release formaldehyde in utero after administration by inhalation or by the dermal or oral route that a second teratogenicity study be conducted on it.

3 That it be established whether humans are likely to be exposed to the condensation product of glycerin and formaldehyde in vivo. If so, additional studies should be conducted by routes most appropriate to human exposure to determine the breadth of the embryotoxic and teratogenic zones.

REFERENCES

1 Connors, TA. Cytotoxic agents in teratogenic research. In *Teratology Trends and Applications*, edited by CL Berry and DE Poswillo, pp. 49–79. New York: Springer-Verlag (1975).

2 Pushkina, NN, VA Gofmekler, and GN Klevetsova. Changes in content of ascorbic acid and nucleic acids produced by benzene and formaldehyde. *Bull. Exp. Biol. Med.* 66:868–870 (1968).

3 Ritter, EJ, WJ Scott, and JG Wilson. Teratogenesis and inhibition of DNA synthesis induced in rat embryos by cytosine arabinoside. *Teratology* 4(1): 7–14 (1971).

4 Gofmekler, VA. Effect on embryonic development of benzene and formaldehyde in inhalation experiments. *Hyg. Sanit.* 33(3):327–332 (1968).

5 Gofmekler, VA, and TI Bonashevskaya. Experimental studies of teratogenic properties of formaldehyde, based on pathological investigations. *Hyg. Sanit.* 34(5):266–268 (1969).

6 Guseva, VA. Study of the gonadotropic effect in male rats under the effect of formaldehyde administered simultaneously with air and water. *Hyg. Sanit.* 37(10):102–103 (1972).

7 Kalmykova, TP. Cytogenetic effect of formalin on somatic and germ cells of animals. *Veterinariia* 11:67–69 (1979).

8 Ranstroem, S. Stress and pregnancy. *Acta Pathol. Microbiol. Scand.* Suppl. 111:113–114 (1956).

9 Isaacson, RJ, and AP Chaudhry. Cleft palate induction in strain A mice with cortisone. *Anat. Rec.* 142(4):479–484 (1962).

10 Hurni, H, and H Ohder. Reproduction study with formaldehyde and hexamethylenetetramine in beagle dogs. *Food Cosmet. Toxicol.* 11:459–462 (1973).

11 Marks, TA, WC Worthy, and RE Staples. Influence of formaldehyde and Sonacide (potentiated acid glutaraldehyde) on embryo and fetal development in mice. *Teratology* 22(1):51–58 (1980).

12 Kimura, ET, DM Ebert, and PW Dodge. Acute toxicity and limits of solvent residue for sixteen organic solvents. *Toxicol. Appl. Pharmacol.* 19:699–704 (1971).

13 Kloecking, HP, and M Richter. Toxicological studies of the harmfulness of methanol in drinking water. *Fortschr. Wasserchem. Ihrer. Grenzgeb.* 14: 189–193 (1972).

14 Della Porta, G, JR Cabral, and G Parmiani. Transplacental toxicity and carcinogenesis studies in rats with hexamethylenetetramine. *Tumori* 56: 325–334 (1970).

15 Natvig, H, J Andersen, and EW Rasmussen. A contribution to the toxicological evaluation of hexamethylenetetramine. *Food Cosmet. Toxicol.* 9:491–500 (1971).

16 Jackson, H, and AW Craig. Antifertility action and metabolism of hexamethylphosphoramide. *Nature* 212:86–87 (1966).

17 Kimbrough, R, and TB Gaines. Toxicity of hexamethylphosphoramide in rats. *Nature* 211:146–147 (1966).

18 Shott, LD, AB Bořkovec, and WA Knapp. Toxicology of hexamethylphosphoric triamide in rats and rabbits. *Toxicol. Appl. Pharmacol.* 18:499–506 (1971).

19 Giavini, E, and M Prati. Cardiovascular malformations experimentally induced in the rat. *Acta Anat.* 106:203–211 (1980).

20 Aliverti, V, L Bonanomi, P DiTrapani, E Giavini, VG Leone, and L Mariani. Effects of glycerol formal on the embryonic development of rats. *Arch. Sci. Biol.* 61:89–95 (1977).

21 Giavini, E, VG Leone, and M Prati. Teratogenesis of glycerol formal: Determination of the period of dysmorphogenic action in rats. *Acta Embryol. Exp.* 3:377–378 (1978).

Part Two

Toxicology

Chapter 7

Formaldehyde Effects in the C3H/10T½ Cell Transformation Assay

Craig J. Boreiko

Daniel L. Ragan

Numerous studies have been conducted over the past several decades in an effort to define the genotoxic activity of formaldehyde (1). Formaldehyde is known to be mutagenic for organisms such as *Drosophila*, fungi, and some bacteria, and more recent studies have found formaldehyde to be capable of inducing single-stranded DNA breaks (2), DNA-protein cross-links (3), and mutations (4) in yeast. The genotoxic activity of formaldehyde has also been examined in cultured mammalian cells. Formaldehyde will induce sister chromatid exchanges in Chinese hamster ovary cells (5), but it does not appear to be mutagenic for these cells (6). Formaldehyde will induce unscheduled DNA synthesis in Hela cells (7), but it fails to induce a response in monkey kidney cells (8). The significance of these studies is far from being clear. The sensitivity of test organisms to formaldehyde genotoxicity is extremely variable and can depend on such diverse parameters as the age, sex, physiological state, and nutritional status of the organism under examination. For example, while orally administered formaldehyde is known to cause mutations in *Drosophila*, this mutagenic activity is restricted to male fruit fly larva. More specifically, it is mutagenic for a specific development stage of the spermatocyte, and even then the induction of these mutations requires the presence of adenosine, adenylic acid, or RNA in the culture medium (1). Formaldehyde is thus a chemical with genotoxic potential, but this potential is realized only under extremely select conditions. Meaningful extrapolation of these results to humans has not been possible.

The need to understand the genotoxic potential of formaldehyde has been emphasized by the preliminary results of a recent inhalation study sponsored by the Chemical Industry Institute of Toxicology (9). In this study, the exposure of rats to 15 ppm formaldehyde vapor produced numerous squamous cell carcinomas of the nasal cavity after 18 months of exposure. The induction of carcinomas by formaldehyde through the 24-month sacrifice point of this study followed a very sharp dose response, and only weak carcinogenic activity was observed in mice exposed to formaldehyde at the same time as rats (see Kerns et al., Chap. 11, this volume).

To provide additional information that would facilitate efforts to assess formaldehyde's carcinogenic potential, we have studied the effects of formaldehyde on $C3H/10T\frac{1}{2}$ cells. $C3H/10T\frac{1}{2}$ C18 is a cell line of mouse embryo fibroblasts isolated from the embryos of C3H mice (10). The cells are nontumorigenic when injected into syngeneic animals, and when grown in culture, they display density-dependent restrictions on proliferation. When small numbers of $C3H/10T\frac{1}{2}$ cells are plated into a cell culture dish, they will grow to form a confluent monolayer across the bottom of the dish and then cease cell division. If, however, cells are plated, exposed to a carcinogen, and incubated for several weeks after confluence has been achieved, discrete multilayered foci of transformed cells become apparent against a background monolayer of nontransformed cells (11). The cells within these foci have lost their density-dependent controls on growth, and if isolated, grown in mass culture, and injected into syngeneic animals, they will frequently grow to form fibrosarcomas.

$C3H/10T\frac{1}{2}$ cells have been found to be transformable by diverse classes of carcinogens and have had widespread applications for the study of chemical carcinogenesis (see reference 12 for a review). Furthermore, the process by which these cells are oncogenically transformed frequently seems to mimic the multistep nature of in vivo carcinogenesis. As has been most clearly demonstrated on mouse skin (13), in vivo tumor development can be divided into at least two discrete steps. The first step, initiation, is irreversible and can be accomplished by numerous carcinogens. Cells initiated by carcinogens need not result in malignancies. The second step, promotion, is reversible and can be accomplished by repeated treatments with agents known as tumor promoters, which facilitate the development of malignancy in cells that have been altered by previous contact with an initiating agent. Tumor promoters are typically nonmutagenic and are not of themselves carcinogenic. Recent studies have found that the process of $C3H/10T\frac{1}{2}$ transformation can be initiated by X-rays, ultraviolet light, and polycyclic aromatic hydrocarbons to produce few, if any, transformed foci (12). If, however, initiated $C3H/10T\frac{1}{2}$ cells are treated repeatedly with the potent tumor promoter 12-O-tetradecanoyl-phorbol-13-acetate (TPA), numerous transformed foci will result (12). The C3H/10T$\frac{1}{2}$ cell culture system thus has applications for the study and detection of complete carcinogens, initiating agents, and tumor promoters. The versatility of the C3H/10T$\frac{1}{2}$ transformation system makes it ideal for the study of formaldehyde.

CYTOTOXICITY STUDIES

As a prelude to transformation studies, experiments were conducted to determine the cytotoxic effects of formaldehyde. All studies were conducted using commercially available 37% w/w formaldehyde solutions (Fisher Scientific). Cell culture studies utilized C3H/10T$\frac{1}{2}$ cells between passages 6 and 12. Stock cultures were maintained in accordance with the procedures of Reznikoff et al. (10) in a medium of BME (Gibco) supplemented with 10% heat-inactivated fetal calf serum (Reheiss). The cytotoxicity of formaldehyde was assessed by the plating efficiency reduction method of Reznikoff et al. (11). Dishes containing 2×10^2 to 2×10^4 cells were treated for 24 h with commercial formaldehyde (Fisher Scientific) diluted with normal Dulbecco's phosphate-buffered saline (Gibco). Negative controls were treated with phosphate-buffered saline. Following treatment, the dishes were washed with phosphate-buffered saline, provided with fresh medium, and incubated 7–9 days to allow colonies of cells to form. The dishes were then washed with 0.9% NaCl fixed in absolute methanol and stained with 5% Giemsa. The colonies that had formed were counted, a decrease in colony number being interpreted as an indication of cytotoxicity. Results were then plotted as "Surviving Fraction" (number of colonies per dish at a given treatment concentration divided by the number of colonies per dish in mock treated controls) vs. concentration. As is shown in Fig. 1, a 24-h exposure of C3H/10T$\frac{1}{2}$ cells to formaldehyde concentrations ranging from 0.1 μg/ml to 4.0 μg/ml produced a steep dose response for cytotoxicity. The dose response curve is biphasic, with an LD_{50} concentration between 0.5 μg/ml and 1.0 μg/ml.

A series of experiments were also performed in order to determine the time course of formaldehyde cytotoxicity. Cells were exposed to formaldehyde concentrations of 1.0, 2.5, or 25.0 μg/ml for periods of time ranging from 15 min to 48 h. The results of these studies are shown in Fig. 2. The concentration $\times$ time product required for 50 percent cytotoxicity was found to be relatively constant (about 20–30 μg/ml/h).

TRANSFORMATION STUDIES

Experiments were conducted to determine formaldehyde's ability to transform C3H/10T$\frac{1}{2}$ cells. Cells were treated for 24 h with concentrations of formaldehyde that ranged from 0.1 μg/ml to 2.5 μg/ml. The choice of a 24-h treatment

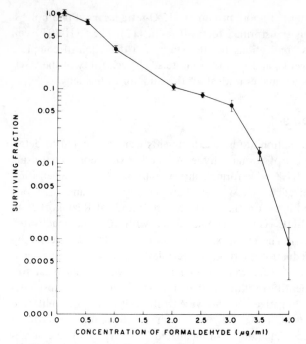

Figure 1 The cytotoxic effects of formaldehyde on C3H/10T$\frac{1}{2}$ cells. Each data point is generally the average of two or more experiments with four to six dishes per treatment group. The vertical bars indicate the standard error of the mean.

period is usual in the C3H/10T$\frac{1}{2}$ assay since it allows ample time for the metabolism of compounds and also results in the exposure of cells to test compounds for a period of time that exceeds one full cell cycle. The range of concentrations examined was selected to include nontoxic and moderately toxic (approximately 10 percent survival) doses.

Transformation assays were conducted following a modification of the basic protocols of Reznikoff et al. (11). Briefly, 2000 cells were plated in 60-mm dishes with complete medium containing 10% fetal calf serum and treated with formaldehyde one day later for a period of 24 h (24–36 dishes per concentration). Dishes treated with 2.5 µg/ml formaldehyde received 2 X 10^4 cells to compensate for cell killing. Additional low density cultures were treated in order to simultaneously assess cytotoxicity. The medium in the dishes was then renewed twice a week until the dishes were confluent and once a week thereafter. A portion of each treatment group (12–24 dishes) received medium containing 0.1 µg/ml TPA with every change of medium from day 5 onward. From day 22 on, all groups received medium in which the concentration of fetal calf serum had been reduced to 5%, as described by Bertram (14). Six weeks after treatment, the cultures were fixed, stained, and scored for the presence of type II and type III foci. Transformation frequencies were then calculated using the number of all type II and type III foci observed and the number of survivors, as estimated from the cytotoxicity dishes of each experiment.

The results of these transformation studies are graphically depicted in Fig. 3. The exposure of $C3H/10T\frac{1}{2}$ cells to formaldehyde alone did not result in significant rates of transformation. A single transformed focus was observed in 1 of 33 dishes treated with 2.5 $\mu g/ml$ formaldehyde. No foci were observed in phosphate-buffered saline-treated dishes. Each experiment also included a small number of dishes treated with the carcinogen 3-methylcholanthrene as a positive control (data not shown). Foci were observed in 22 of 27 dishes treated with 3-methylcholanthrene, indicating that the cells in these experiments were susceptible to transformation.

When the treatment of cells with formaldehyde was followed by continuous treatment with the tumor promoter TPA, numerous transformed foci were produced. The frequency of focus production increased in a dose-dependent fashion from 0.1 $\mu g/ml$ to 1.0 $\mu g/ml$ of formaldehyde and then decreased at a formaldehyde concentration of 2.5 $\mu g/ml$. No foci were observed in dishes

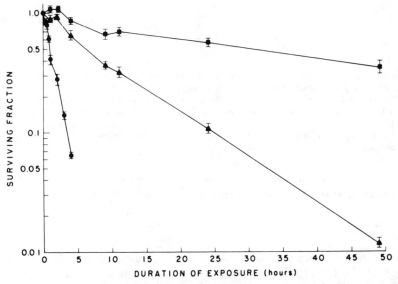

Figure 2 The time course of formaldehyde cytotoxicity. Cell culture conditions, addition of formaldehyde, and staining of dishes are described in text. Cells were exposed to 1.0 (■), 2.5 (▲), or 25.0 (●) $\mu g/ml$ formaldehyde. Following the indicated duration of exposure to formaldehyde, the dishes were washed with Dulbecco's phosphate-buffered saline, complete medium was added, and the dishes were incubated as described previously. Cytotoxicity is expressed as the surviving fraction relative to mock-treated duplicate cultures. The results are the average of three experiments, each having four to six dishes per time point, for each concentration of formaldehyde. Standard error of the mean is indicated by the vertical bars.

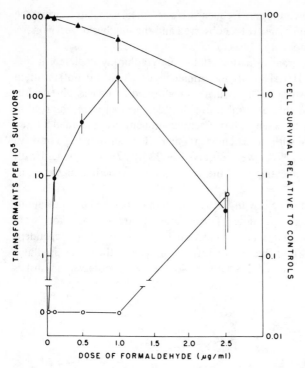

Figure 3 The production of transformed foci by formaldehyde. Conditions for the cul-
turing of cells and formaldehyde treatment were as described for cytotoxicity studies.
Results presented are the mean of four experiments and show the dose response for the
observed cytotoxicity of formaldehyde (▲), the frequency of focus production following
treatment with formaldehyde alone (o), and the frequency of focus formation when formaldehyde treatment is followed by the application of tumor promoters (●). The standard
error of the mean is indicated by the vertical bars.

treated with saline and then continuously treated with TPA. Representative
dishes from these transformation experiments are pictured in Fig. 4.

DISCUSSION

These experiments indicate that although aqueous formaldehyde solutions are
quite toxic for $C3H/10T\frac{1}{2}$ cells, a single treatment of the cells with formaldehyde is not sufficient to produce the transformed state. If, however, treatment
with formaldehyde is followed by the continuous application of the tumor promoter TPA, numerous transformed foci are produced. Formaldehyde is thus an
initiating agent for transformation in a manner quite similar to that of ultraviolet
light (12).

Certain technical aspects of this work deserve consideration. These studies
utilized aqueous formaldehyde prepared from commercial formaldehyde solu-

tions. These commercial solutions contain a significant concentration of methanol (10%). Furthermore, the formaldehyde in such solutions is present in a hydrated form (methylene glycol) and reacts reversibly with methanol to form hemiacetals (15). These facts raise questions as to the nature of the reactive species responsible for the initiation of transformation. Other studies in this laboratory have found that concentrations of methanol up to 100 mg/ml do not possess significant initiating or transforming activity for $C3H/10T\frac{1}{2}$ cells (unpublished data). This concentration is 10^5 times higher than that present as a contaminant in these studies. The initiation of transformation by formaldehyde is thus not a direct result of the presence of methanol. The presence of formaldehyde in the form of methylene glycol is not of concern either, since this is the form formaldehyde vapor would adopt on dissolving in the aqueous environment within living tissues or cells (15). The possibility that hemiacetals are involved in the initiating activity detected by the $C3H/10T\frac{1}{2}$ system cannot be discounted. Under the conditions of temperature, pH, and concentration used to treat the cells, however, such compounds should rapidly dissociate to methylene glycol and methanol (15). Even accepting the probability that the initiating activity of formaldehyde solutions is due to the presence of hydrated formaldehyde monomer, one cannot be sure that this effect is a direct one. Other studies on the mutagenic activity of formaldehyde have suggested that the mutagenic

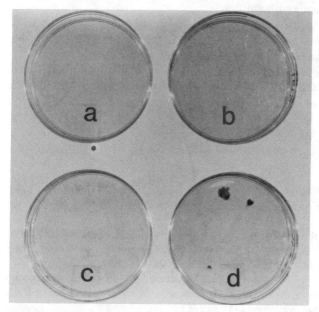

Figure 4 Representative dishes from transformation studies. The dishes depict the typical result for dishes treated with *a*) phosphate-buffered saline, *b*) phosphate-buffered saline followed by continuous treatment with TPA, *c*) 1.0 µg/ml of formaldehyde, and *d*) 1.0 µg/ml of formaldehyde followed by continuous treatment with TPA.

activity is produced by the reaction of formaldehyde with select components of media (1). The concentrations of formaldehyde used in these studies require hours to exert their cytotoxic effects, a finding that is surprising in view of the highly reactive nature of formaldehyde (15). While numerous explanations for this delayed cytotoxicity can be offered, the possibility that formaldehyde-initiating activity is mediated by the products of its reaction with the constituents of cell culture medium should be considered. In addition, if biologically active forms of formaldehyde are able to persist in vivo, then formaldehyde may be capable of exerting genotoxic effects at tissue sites distant from the respiratory tract. A recent study suggests this can occur; mice exposed to formaldehyde vapor are reported to have an increased frequency of sister chromatid exchanges in bone marrow (see Brusick, Chap. 8, this volume).

The finding that formaldehyde is a pure initiating agent for $C3H/10T\frac{1}{2}$ mouse embryo cells can be interpreted as having implications for the recent inhalation study that found formaldehyde to be carcinogenic for rats (9). First, it suggests that formaldehyde carcinogenesis may, in certain instances, be a multistage process. The existence of promotional factors in the induction of rat neoplasms should thus be considered. In view of the extreme species and tissue specificity for initiation and promotion in vivo (13), however, comparisons between the transformation of $C3H/10T\frac{1}{2}$ cells and in vivo formaldehyde carcinogenesis must be made with caution.

This work also has two important practical implications. First, the $C3H/10T\frac{1}{2}$ system has been found to be responsive to formaldehyde, thereby establishing the utility of the system for the study of chemicals that may interact with biological organisms by mechanisms similar to that of formaldehyde. Second, because the $C3H/10T\frac{1}{2}$ system is responsive to formaldehyde, the cells provide a model system for the study of formaldehyde's biochemical mechanism of action.

REFERENCES

1 Auerbach, C, M Moutschen-Dahmen, and J Moutschen. Genetic and cyto-genetical effects of formaldehyde and related compounds. *Mutat. Res.* 39: 317–362 (1977).

2 Magana-Schwencke, N, B Ekert, and E Moustacchi. Biochemical analysis of damage induced in yeast by formaldehyde. I. Induction of single strand breaks in DNA and their repair. *Mutat. Res.* 50:181–193 (1978).

3 Magana-Schwencke, N, and B Ekert. Biochemical analysis of damage induced in yeast by formaldehyde. II. Induction of cross-links between DNA and protein. *Mutat. Res.* 51:11–19 (1978).

4 Chanet, R, and RC von Borstel. Genetic effects of formaldehyde in yeast. III. Nuclear and cytoplasmic mutagenic effects. *Mutat. Res.* 62:239–253 (1979).

5 Obe, G, and B Beek. Mutagenic activity of aldehydes. *Drug Alcohol Depend.* 4:91–94 (1979).

6 Hsie, AW, J O'Neill, JR San Sebastian, DB Couch, PA Brimer, WNC Sun, JC Fuscoe, NL Forbes, R Machanoff, JC Riddle, and MH Hsie. Quantitative mammalian cell genetic toxicology: Study of the cytotoxicity and mutagenicity of seventy individual environmental agents related to energy technologies and three subfractions of a crude synthetic oil in the CHO/HGPRT system. EPA Report 600/9-78-027, pp. 293–315 (1979).

7 Martin, CN, AC Mcdermid, and RC Garner. Testing of known carcinogens and noncarcinogens for their ability to induce unscheduled DNA synthesis in HeLa cells. *Cancer Res.* 38:2621–2627 (1978).

8 Nocentini, S, G Moreno, and J Coppey. Survival, DNA synthesis and ribosomal RNA transcription in monkey kidney cells treated by formaldehyde. *Mutat. Res.* 70:231–240 (1980).

9 Swenberg, JA, WD Kerns, RI Mitchell, EJ Gralla, and KL Pavkov. Induction of squamous cell carcinomas of the rat nasal cavity by inhalation exposure to formaldehyde vapor. *Cancer Res.* 40:3398–3402 (1980).

10 Reznikoff, CA, DW Brankow, and C Heidelberger. Establishment and characterization of a cloned line of C3H mouse embryo cells sensitive to post-confluence inhibition of cell division. *Cancer Res.* 33:3231–3238 (1973).

11 Reznikoff, CA, JS Bertram, DW Brankow, and C Heidelberger. Quantitative and qualitative studies of chemical transformation of clones of C3H mouse embryo cells sensitive to post-confluence inhibition of cell division. *Cancer Res.* 33:3239–3249 (1973).

12 Mondal, S. The $C3H/10T\frac{1}{2}$ Cl 8 mouse embryo cell line: Its use for the study of carcinogenesis and tumor promotion in cell culture. In *Advances in Environmental Toxicology. Vol. 1. Mammalian Cell Transformation by Chemical Carcinogens*, edited by MA Mehlman, NK Mishra, and V Dunkel. Princeton Junction, NJ: Senate, pp. 181–211 (1980).

13 Scribner, JD, and R Süss. Tumor initiation and promotion. *Int. Rev. Exp. Pathol.* 18:137–198 (1978).

14 Bertram, JS. Effects of serum concentrations on the expression of carcinogen-induced transformation in the $C3H/10T\frac{1}{2}$ Cl 8 cell line. *Cancer Res.* 37:514–523 (1977).

15 Walker, JF. *Formaldehyde*, 3d ed. New York: Krieger (1975).

Chapter 8

Genetic and Transforming Activity of Formaldehyde

David J. Brusick

A genetic evaluation of formaldehyde was designed around a two-phased approach as outlined in Table 1. The first phase consisted of in vitro evaluation of formaldehyde with a battery of submammalian and in vitro mammalian test systems to evaluate genotoxic activity of formaldehyde. The in vitro test battery consisted of the Ames *Salmonella* reverse mutation assay, a mouse lymphoma forward mutation assay employing the TK locus, tests measuring the induction of sister chromatid exchange (SCE) and chromosome aberrations in Chinese hamster ovary (CHO) cells, and a cell transformation test using Balb/c3T3 cells.

The second phase is presently being conducted and consists of an evaluation of the genotoxic activity of formaldehyde in vivo. The in vivo battery of tests is being performed on mice exposed to formaldehyde vapors and includes determination of SCE, chromosome aberrations, and recessive coat color spot mutations. In this report results from some of the SCE studies are presented.

The study was sponsored by the Formaldehyde Institute and performed at Litton Bionetics, Inc., Department of Molecular Toxicology, Kensington, Maryland.

Table 1   Two-Phased Analysis of Formaldehyde Genotoxicity

Phase I—in vitro
  Ames test [agar incorporation technique)
  Mouse lymphoma forward mutation assay
  Balb/c3T3 cell transformation test
  Analysis for sister chromatid exchange (SCE) and chromosome aberrations in Chinese
    hamster ovary cells
Phase II—in vivo (inhalation exposure of mice)
  Analysis of somatic mutation induction in embryonic melanocytes
  Analysis for SCE in bone marrow cells
  Analysis for chromosome aberrations in bone marrow cells
Planned Phase II—exposure levels
  High   25 ppm
  Medium   12 ppm
  Low   6 ppm

MATERIAL AND METHODS

In Vitro

The submammalian and mammalian in vitro portions of this study were conducted according to the standard protocols used by Litton Bionetics, Inc. (1).

All short-term studies were conducted with various concentrations of formaldehyde prepared by dissolving solid paraformaldehyde (92 percent) in water. Preliminary toxicity tests were conducted for each study to determine the appropriate dose range. The solubility of paraformaldehyde is low, and the concentrations reported in the results are based on applied concentration of paraformaldehyde. Dose selections were based on evidence of cytotoxicity, and the ranges varied as a function of exposure time and the presence or absence of activator (Arochlor 1254-induced rat liver S9 mix).

In Vivo

Test Animals Adult male and female CD-1 mice were used on the SCE study and for the chromosome aberration study. Adult, Charles River C57B1/6J female mice and adult male T-strain mice from the Litton Bionetics breeding colony were used in the spot test.

All animals were quarantined at least one week prior to use in the study. Female mice were housed up to 15 per plastic box cage with AB-SORD-DRI bedding. Certified Purina Laboratory Chow and Breeders Chow were offered ad libitum with water. Light was provided on a 12-h light/dark cycle. Food and water were removed during the inhalation exposures.

Animals were assigned to study groups at random and were identified by ear tag. Dose or treatment groups were identified by cage cards. Animals were housed individually in wire cages during exposures in inhalation chambers.

Exposure Chambers and Test Substance Generation The animals were exposed to formaldehyde vapors or to control air in 8.2 cubic meter stainless steel and glass exposure chambers operated under dynamic conditions with negative pressure. Total airflow for both test and control chambers was 35 cubic feet per minute.

Formaldehyde vapor was generated by heating approximately 50 g of paraformaldehyde powder (92 percent) to a temperature of $55°$-$90°C$ in a three-necked flask. The concentrated vapor was moved into the dilution airstream by passing dry air through the flask. Concentrations were controlled by adjusting the airflow through the flask and the flask temperature. The dilution air was filtered through HEPA filters and activated charcoal, cooled to $16.1°C$, and humidified to 40–60 percent R.H.

The chamber concentrations were monitored at least every 2 h using the National Institute of Occupational Health and Safety modification (2) of MacDonald's field method (3).

Exposure Calibrations A standard curve was constructed from analysis of formaldehyde solutions and analyzed spectrophotometrically as described above.

The effective duration of the exposure was 6 h daily. There was an additional 20-min period at the end of the exposure to allow for the chamber to reach equilibrium with room air. This time was determined by the equation

$$t_{99} = 4.6 \, \frac{\text{chamber volume}}{\text{chamber airflow}}$$

Test Design *Sister Chromatid Exchange* The following procedure was used for detection of SCE: Groups of five male and five female CD-1 mice were exposed to target concentrations of 0, 6, or 12 ppm formaldehyde vapor 6 h/day for 5 days or 25 ppm 6 h/day for 4 days.

For detection of SCE, a table of bromodeoxyuridine (about 55 mg) was implanted subcutaneously within 2 h after the end of inhalation or, in the case of positive controls, immediately before treatment with cyclophosphamide, and the animals were killed 22 or 25 h later.

About 3 h prior to kill, the animals were injected intraperitoneally with 4 mg/kg of colchicine. The animals were killed with CO_2. The adhering soft tissue and epiphyses of both tibias were removed according to the method of Legator et al. (4). The marrow was aspirated from the bone and transferred to Hanks balanced salt solution. The marrow button was collected by centrifugation and then resuspended in 0.075 M KCl for about 10 min. The centrifugation was repeated and the pellet resuspended in fixative (methanol:acetic acid, 3:1). After a further wash in fixative, cells were stored in fixative at 4°C for up to 3 weeks before slides were prepared. Cells in fixative were dropped onto glass slides and air dried.

Slides for SCE analysis were stained for 10 min with Hoechst Stain No. 33258 (5 g/ml) in phosphate buffer (pH 6.8), mounted in the same buffer, and exposed to ultraviolet (UV) light from a 215 watt "black light" tube at 60°C for the amount of time required for sister chromatid differentiation. Following UV exposure, the slides were stained with 10% Giemsa for 10 min and air dried. Second-division cells (M2 cells) were scored for the frequency of SCE per cell and per chromosome.

A student's t-test was employed using results from individual animals as data points. An increase of $p < 0.01$ was considered significant.

Chromosome Aberrations For detection of chromosome aberrations, male and female CD-1 mice were exposed to target concentrations of 0, 6, or 12 ppm formaldehyde 6 h/day for 5 days or 25 ppm 6 h/day for 4 days. The mice were killed 6 h after the last inhalation exposure. Animals were injected intraperitoneally with colchicine and the bone marrow was collected and treated as described for SCE. Slides were stained with 5% Giemsa and scored for chromosome aberration frequencies. Fifty cells were scored per animal. For control of bias, all slides were coded prior to technician scoring.

Chromosomes in the metaphase cells were screened for gaps, breaks, fragmentation, deletions, rearrangements, and polyploidy. The estimated number of breaks involved in production of the different types of aberrations was observed. Gaps were not counted as significant aberrations. The frequency of cells with more than one aberration was observed. Statistical analysis was done with the student t-test.

Mouse Spot Test The objective of this study was to detect induced specific-locus somatic mutations in vivo using specially derived strains of mice. The mutations were initially induced in melanocyte percursor cells in developing embryos that were heterozygous at five specific coat color loci. Melanocyte precursor cells carrying a mutation develop into clones of "mutant" melanocytes that can be readily recognized as coat color mosaic patches. Since each developing embryo contains approximately 150-200 melanocyte precursor cells, a relatively small number of animals is required to provide reliable data in this in vivo somatic mutation assay (5-7).

The method consists of exposing pregnant females to formaldehyde via inhalation 6 h/day on days 9-11 of pregnancy and subsequently examining the young after birth for mosaic patches on their coats. The negative control group was exposed to filtered air. The positive control group was treated with ethylnitrosourea (50 mg/kg intraperitoneally) on day 10 of pregnancy. The newborns were scored for mosaic coat color spots on day 12 and again on day 24 post partum. White midventral spots were considered to represent melanocyte toxicity and not mutational events. Incidence of coat color spots in the treated and the positive control groups was compared against the negative control to determine the p-value according to Fisher's exact test.

RESULTS

In Vitro Assays

Table 2 summarizes the results of formaldehyde in the phase I test battery. Three assays were positive—mouse lymphoma, SCE, and cell transformation in Balb/c3T3. The Ames assay and chromosome aberration tests were negative.

The Ames *Salmonella* reverse mutation assay results for formaldehyde are presented in Table 3. Formaldehyde was negative up to 1000 μg/plate both with and without metabolic (S9) activation. Recent National Toxicology Program results show a positive formaldehyde response in a preincubation version of the Ames test.

The mouse lymphoma assay was positive both with and without metabolic activation. Figure 1 illustrates mutation and relative growth rates of nonactivated mouse lymphoma cells versus formaldehyde concentration. The results with S9 activation are shown in Fig. 2. The presence of S9 mix lowered the minimum effective mutagenic concentration of formaldehyde in mouse lymphoma assay from 7.5 μg/ml to 1.9 μg/ml.

The assay for SCE in CHO cells was positive with and without metabolic activation. As in the mouse lymphoma assay, S9 activation lowered the minimum effective concentration for the induction of SCE. Table 4 gives frequencies

Table 2   Phase I Summary Table of Results

Test	S9 required	Response	Minimum effective concentration (MEC) or maximum concentration tested (MCT)
Ames test	No	Negative	1000 μg/plate (MCT) both with and without S9 mix
Mouse lymphoma assay	No	Positive	1.9 μg/ml (MEC) with S9 mix 7.5 μg/ml (MEC) without S9 mix
CHO SCE assay	No	Positive	0.5 μg/ml (MEC) with S9 mix 1.0 μg/ml (MEC) without S9 mix
CHO aberration assay	No	Negative	3.0 μg/ml (MCT) with S9 mix 4.0 μg/ml (MCT) without S9 mix
Cell transformation assay (Balb/c3T3)	No	Positive	Range is 0.5 through 2.5 μg/ml

CHO = Chinese hamster ovary cell line; SCE = sister chromatid exchange.

Table 3 Results of Formaldehyde in the Ames *Salmonella* Reverse Mutation Assay

Concentration (µg/plate)	Revertants per plate									
	TA-1535		TA-1537		TA-1538		TA-98		TA-100	
	−S9	+S9	−S9	+S9	−S9	+S9	−S9	+S9	−S9	+S9
Solvent	9	13	35	34	11	19	21	24	155	135
0.5	10	15	32	36	19	19	26	24	121	108
1.0	9	14	34	44	17	11	20	34	131	160
10.0	9	13	43	45	23	22	19	27	138	166
100.0	7	14	33	45	19	9	33	38	144	144
500.0	0	1	34	32	15	12	7	23	46	34
1000.0	0	0	0	3	0	6	0	0	8	0
Positive control*	806	306	305	231	1732	1371	1001	1575	1696	1990

S9 used was Lot #IRL-CH 172. Solvent used was 50% dimethysulfoxide (DMSO).

*

	−S9	+S9
TA-1535	Sodium azide 10 µg/plate	2-Aminoanthracene 2.5 µg/plate
TA-1537	9-Aminoacridine 50 µg/plate	2-Aminoanthracene 2.5 µg/plate
TA-1538	2-Nitrofluorene 10 µg/plate	2-Aminoanthracene 2.5 µg/plate
TA-98	2-Nitrofluorene 10 µg/plate	2-Aminoanthracene 2.5 µg/plate
TA-100	Sodium azide 10 µg/plate	2-Aminoanthracene 2.5 µg/plate

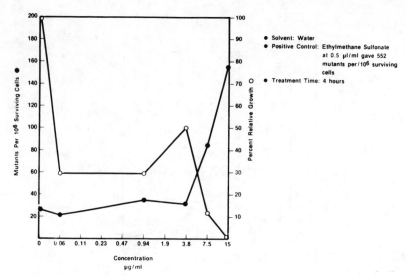

Figure 1 Mutagenic activity of formaldehyde at the TK locus in mouse lymphoma (L5178Y) cells. Nonactivation test conditions.

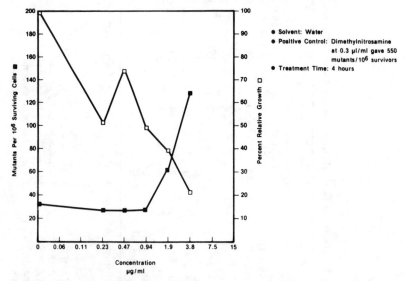

Figure 2 Mutagenic activity of formaldehyde at the TK locus in mouse lymphoma (L5178Y) cells. Activation test conditions.

78

Table 4 Sister Chromatid Exchange (SCE) Frequencies in Cells Exposed to Formaldehyde

Treatment	Dose	No. of chromosomes	No. of SCE	Nonactivation		
				SCE/Chromosome ± SE	SCE/Cell	
Neg. control (medium)		760	166	0.218 ± 0.017	8.74	
Vehicle control	1.0 ml/flask	752	225	0.299 ± 0.020	11.97	
Positive control: ethylmethane sulfonate	0.5 µl/ml	738	1338	1.813 ± 0.050*	72.52*	
Test compound						
Formaldehyde	1.0 µg/ml	733	313	0.427 ± 0.024*	17.08*	
Formaldehyde	2.0 µg/ml	741	257	0.347 ± 0.022	13.87	
Formaldehyde	3.0 µg/ml	740	280	0.378 ± 0.023*	15.14*	
Formaldehyde	4.0 µg/ml	736	343	0.466 ± 0.025*	18.64*	

*Significantly greater than vehicle control, $p < 0.01$ (Student's t-test).

of exchanges in nonactivated CHO cells. Nonactivated cells showed significant SCE increases at concentrations of 1 μg/ml, 3 μg/ml, and 4 μg/ml. There was no dose-response relationship. With metabolic activation (Table 5), concentrations of 0.5 μg/ml and 2 μg/ml produced significant increases of SCE, while 1 μg/ml did not. Higher concentrations were cytotoxic. As in the nonactivated system, no dose-response relationship was observed with activation.

No chromosome aberrations were observed at concentrations up to 3 or 4 μg/ml in CHO cells, with or without S9 mix, respectively. Concentrations higher than this were completely cytotoxic and inhibited mitotic division (data not shown).

Formaldehyde transformed mouse cells in the absence of an exogenous metabolic system. Cell transformation of Balb/c3T3 cells exposed over a range of 0.5 to 2.5 μg/ml produced a dose-response effect starting from approximately one focus per plate up to a mean of four foci per plate. The dose-response curve is illustrated in Fig. 3.

In Vivo Assays

The preliminary results of the SCE assay on mice exposed to formaldehyde vapor are summarized in Table 6. Male mice exposed up to a target concentration of 25 ppm formaldehyde 6 h/day for 4 days did not have an increase in SCE. In the high and mid concentration females, there is an elevated level of SCE in comparison to the negative control scores for female mice. There was no effect observed in the low dose female mice. Results for the chromosome aberration analysis and the in vivo mouse somatic mutation assay were negative.

DISCUSSION

Formaldehyde exhibited genotoxic activity in cultured mammalian cells. The activity did not require metabolic activation but appeared to be slightly enhanced in the presence of Arochlor 1254-induced rat liver S9 mix. Formaldehyde induced gene mutation in the mouse lymphoma forward mutation assay as well as SCE in Chinese hamster ovary cells and an increased incidence of cell transformation in Balb/c3T3 cells. The response was dose related in the mouse lymphoma and cell transformation assays but not with SCE. Formaldehyde was not active as a clastogen nor was it active in reverting any of the Salmonella strains when examined in the conventional agar overlay Ames technique. The lack of clastogenic activity is probably associated with the mechanism of DNA interaction. Many of the low molecular weight, highly reactive alkylating agents are not efficient clastogens. High reactivity and rapid binding to agar of the genotoxic molecule may also be responsible for the negative Ames test since there is evidence that formaldehyde is mutagenic in bacteria using a liquid suspension treatment regimen.

Table 5 Sister Chromatid Exchange (SCE) Frequencies in Cells Exposed to Formaldehyde

Treatment	Dose	No. of chromosomes	No. of SCE	Activation SCE/Chromosome ± SE	SCE/Cell
Neg. control (medium)		762	248	0.325 ± 0.021	13.02
Vehicle control	1.0 ml/flask	727	242	0.333 ± 0.021	13.31
Positive control: dimethynitrosamine	0.3 µg/ml	748	829	1.108 ± 0.038*	44.33*
Test compound					
Formaldehyde	0.5 µg/ml	741	350	0.472 ± 0.025*	18.89*
Formaldehyde	1.0 µg/ml	749	304	0.406 ± 0.023	16.23
Formaldehyde	2.0 µg/ml	730	355	0.486 ± 0.026*	19.45*
Formaldehyde	3.0 µg/ml	—	—	Cytotoxic	

*Significantly greater than vehicle control, $p < 0.01$ (Student's t-test).

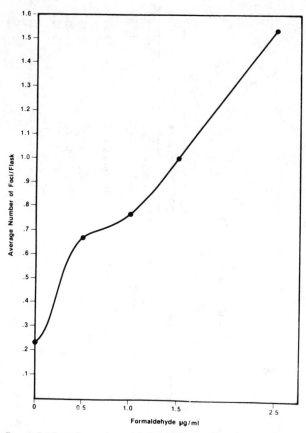

Figure 3 Transformation of Balb/c3T3 cells with formaldehyde.

The level of activity in the in vitro tests was not striking and dose-response curves were not produced in all cases. However, the minimum effective concentration for positive effects ranged from 0.5 to 7.5 µg/ml, indicating genotoxic activity for formaldehyde under these conditions. The apparent threshold of activity of formaldehyde has been described in other in vitro tests and has been attributed to the fact that formaldehyde is a naturally produced intermediary metabolite and low levels of the material may be integrated into normal metabolism. A better picture of the predictability of the phase I in vitro battery will be available when the in vivo studies are completed and evaluated. Preliminary results from the SCE analysis of bone marrow cells from formaldehyde-exposed mice indicate there was evidence for chromosome effects at levels of formaldehyde which were toxic to the mice. At the same levels producing SCE, no chromosome aberrations or somatic mutations were detected.

Table 6 Sister Chromatid Exchange (SCE) Results: Exposure of CD-1 Male and Female Mice to Three Levels of Formaldehyde Vapor

Exposure	Male animal No.	No. of cells scored	No. of SCE	SCE/Cell	Female animal No.	No. of cells scored	No. of SCE	SCE/Cell
High dose*	1090–1094	60	269	4.48	1095–1099	93	1056	11.35§
Mid dose†	0873–0877	100	507	5.07	0898–0902	100	615	6.15§
Low dose†	0868–0872	98	313	3.19	0893–0897	93	300	3.23
Positive control [Cyclophosphamide]	0883–0887	85	1298	15.27§	0908–0912	92	1024	11.13§
Negative control‡	1080–1084	100	330	3.30	1085–1089	60	210	3.50
	0863–0867	100	549	5.49	0888–0892	100	307	3.07
	Combined	200	879	4.40	Combined	160	517	3.23

*Animals exposed 6 h/day × 4 days.
†Animals exposed 6 h/day × 5 days.
‡Animals exposed 6 h/day × 5 days and 6 h/day × 4 days.
§Significantly higher than the negative controls.

SUMMARY

Formaldehyde was evaluated in five in vitro assays: the Ames *Salmonella* reverse mutation assay, the L5178Y mouse lymphoma assay (forward mutation at the TK locus), tests measuring the induction of sister chromatid exchanges (SCE) and chromosome aberrations in Chinese hamster ovary cells (CHO), and a cell transformation test using Balb/c3T3 cells. All in vitro assays except that for cell transformation induction were conducted with and without a hepatic activation system. The Ames test was negative at concentrations up to 1000 μg/plate both with and without activation. The mouse lymphoma assay produced activity at 7.5 μg/ml nonactivated and 1.9 μg/ml activated. The SCE assay produced non-dose-related activity starting at 1 μg/ml nonactivated and 0.5 μg/ml activated. No chromosome aberrations were observed in CHO cells with or without activation. The cell transformation assay produced a dose-response effect over the range of 0.5–2.5 μg/ml.

A second phase of the study was planned to evaluate the genotoxic activity of formaldehyde in vivo by examination of bone marrow cells for SCE in mice exposed to formaldehyde vapor. Increased SCE frequencies were observed in female mice at the mid and high dose levels. However the formaldehyde concentrations were determined to be greater than the targeted concentrations of 12 and 25 ppm. Chromosome aberrations in mouse bone marrow cells were also examined as well as somatic mutation induction in embryonic melanocytes (mouse spot test). A better picture of the predictability of the phase I in vitro battery will be available when the in vivo studies are completed and evaluated.

REFERENCES

1 *Screening Program for the Detection of Mutagens and Carcinogens.* Kensington, MD: Litton Bionetics, Inc. (1979).
2 *NIOSH Manual of Analytical Methods*, Second, Volume 1, Method P and CAM 125.
3 MacDonald, WE, Jr. Formaldehyde in air: A specific field test. *Am. Ind. Health Assoc. J.* 15:217–219 (1954).
4 Legator, MS, et al. Cytogenetic studies in rats of cyclohexalamine, a metabolite of cyclamate. *Science* 165:1139–1140 (1969).
5 Russell, LB. Validation of the in vivo somatic mutation method in the mouse as a prescreen for germinal point mutations. *Arch. Toxicol.* 38:75–85 (1977).
6 Russell, LB, and MH Major. Radiation-induced presumed somatic mutations in the house mouse. *Genetics* 42:161–175 (1975).
7 Fahrig, R. A mammalian spot test: Induction of genetic alterations in pigment cells of mouse embryos with X-rays and chemical mutations. *Mol. Gen. Genet.* 138:309–314 (1975).

Chapter 9

Genetic Toxicology Studies with Formaldehyde and Closely Related Chemicals including Hexamethylphosphoramide (HMPA)

John Ashby

Paul Lefevre

Most of the chapters in this volume are concerned with the toxicological consequences of exposing either bacteria, eukaryotic organisms, or mammals to formaldehyde. The monomer used in these studies is either in the free gaseous state (derived by warming the polymer paraformaldehyde) or as an aqueous solution (Formalin). The relatively high reactivity of formaldehyde to nucleophiles imposes restrictions on its free travel through living tissues, thus most of the toxic responses observed in mammals, including its carcinogenicity, occur at the site of initial contact, usually either the nasal epithelium or the skin. This in turn leads to formaldehyde appearing to be a particularly organ-specific agent, despite the fact that it is potentially able to react with macromolecules of any

The authors wish to thank Mike Penman for assistance with some of the experiments described herein, and John Madden for the art work (Photocall, Manchester).

The unpublished experiments referred to herein were made using the following materials, HMPA (99% pure, BDH Ltd., Poole, Dorset), DMN (Gold Label material, Aldrich Chemical Co. Ltd., Gillingham, Kent), formaldehyde (40% Formalin, BDH Ltd., Poole, Dorset), HMT (Aldrich Chemical Co. Ltd., Gillingham, Kent). The samples of DNPT and DMM were kindly supplied by ICI Limited, Organics Division.

exposed tissue or organ. The present article describes experiments designed to evaluate the genetic toxicology of formaldehyde in tissues that it would not normally reach following exposure by the usual routes, and this was achieved by studying four compounds that decompose to generate formaldehyde in situ. Three of these compounds—dinitrosopentamethylenetetramine (DNPT), hexamethylenetetramine (HMT), and dimorpholinylmethane (DMM)—are solids that readily hydrolyze to formaldehyde under aqueous conditions, and the fourth, hexamethylphosphoramide (HMPA), is a liquid that can be metabolized to formaldehyde via interaction with oxidative enzymes. Differences in the toxicological profiles of these chemicals may cast some light on the potential toxicity of formaldehyde to sites remote from the point of administration, and this may contribute to an understanding of the potential hazard to humans of exposure to formaldehyde.

BIOLOGICAL PROPERTIES OF THE CHEMICALS UNDER CONSIDERATION

The structure of the chemicals discussed in the text are presented in Fig. 1. The estimations of formaldehyde were made according to the method of Nash (1).

Formaldehyde This carcinogen gives a negative response in strains TA98 and TA100 of *Salmonella typhimurium* when tested in the absence of S9 mix in the standard plate test [R Callander, unpublished data (1980)]. In contrast, it gives a positive response in the mouse lymphoma point mutation assay (Fig. 2, ±S9) and in the BHK cell transformation assay (2) (Fig. 3a, ±S9). Similar results have been reported by Brusick (Chap. 8). The mutagenicity of formaldehyde in *Drosophila* has been reviewed by Auerbach (3).

Enzyme-dependent Formaldehyde Generators

Dimethylnitrosamine (DMN) This carcinogen undergoes enzyme-mediated transformation to formaldehyde via α-hydroxylation of one of its methyl groups

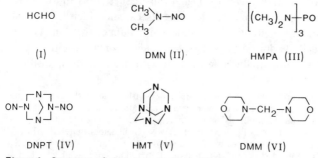

Figure 1 Structure of chemicals mentioned in the text.

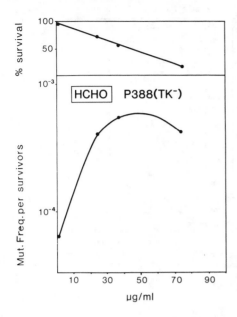

Figure 2 Positive response observed for formaldehyde (−S9) in the mouse lymphoma (P388) TK⁻ mutation assay (21) [Anderson and Cross, unpublished data (1980)].

followed by spontaneous rearrangement of the derived product to formaldehyde and methyldiazonium hydroxide, an alkylating agent (4) (see data shown in Fig. 4). The fact that the carcinogenicity and mutagenicity of DMN are mediated via the derived methyldiazonium compound (as opposed to via the concomitantly derived formaldehyde) excludes DMN for specific consideration as a formaldehyde generator. Nonetheless, it acts as a useful comparison with the enzyme-mediated release of formaldehyde from HMPA.

Hexamethylphosphoramide (HMPA)   This high boiling-point liquid induces squamous cell carcinoma of the nasal epithelium of rats exposed by inhalation (5), and in this respect it closely resembles formaldehyde itself. This resemblance is heightened by the fact that incubation of HMPA with the postmitochondrial fraction of a rat liver homogenate (referred to as S9 mix hereafter) yields formaldehyde (Fig. 4). The activities of HMPA in vitro are also very similar to those of formaldehyde: a negative response in the *Salmonella* assay (6–8) and positive results in mammalian cell test systems (6–8) (e.g., the positive BHK transformation response shown in Fig. 3*b*, +S9). Despite all of these similarities the mode of action of HMPA as a carcinogen has yet to be established unambiguously and it may not be a formaldehyde-mediated effect (8).

The ability of HMPA to dissolve in both aqueous and lipophilic media led us to determine if it would be selectively retained in the nasal epithelium of rats following inhalation. If this were so, then inhalation of apparently low levels of HMPA could lead to its accumulation in this tissue and this might have explained its unusually potent and localized carcinogenicity. The distribution of

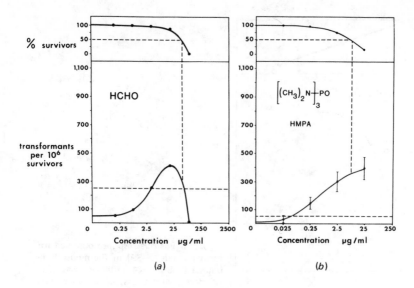

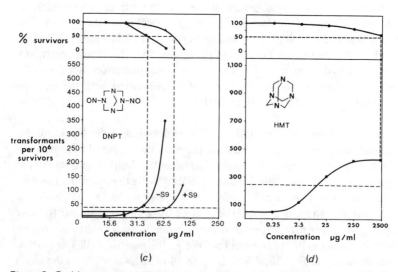

Figure 3 Positive responses observed in the BHK cell-transformation assay (2). *a*) Formaldehyde (I) [−S9; Plesner, unpublished data, (1980)]. *b*) Hexamethylphosphoramide (III) (8) (+S9). *c*) Dinitrosopenthamethylenetetramine (IV) [±S9; Daniel unpublished data, (1980)]. *d*) Hexamethylenetetramine (V) (−S9; Plesner, unpublished data, (1980)]. The methods used have been previously described: the horizontal dotted line represents five times the spontaneous transformation frequency of the clone employed and the vertical dotted line is the LD₅₀ dose level. A positive response is scored when the number of induced transformants per 10⁶ survivors exceeds 5 × background levels at the LD₅₀ dose level.

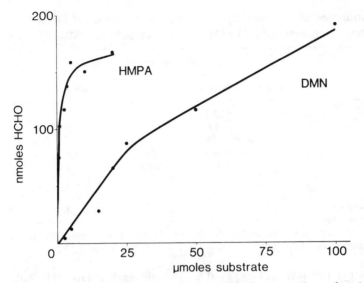

Figure 4 The S9-mediated release of formaldehyde expressed as a function of substrate concentration. Hexamethylphosphoramide (HMPA) and dimethynitrosamine (DMN) were incubated with S9 (the 9000 g postmitochondrial supernatant from the livers of Aroclor 1254-induced Sprague-Dawley rats) and an NADPH generating system at 37°C for 20 min in the presence of semicarbazide. The formaldehyde generated was estimated using the Nash method (1).

radioactivity was studied in the tissues of rats 24 h after the nasal administration of 10 mg of either hexa[^{14}C] methylphosphoramide (342 μCi/mMol, custom synthesis, Radiochemical Centre, Amersham) or [^{14}C] methyl methanesulphonate (98 μCi/mMol, Radiochemical Centre, Amersham). Homogenized tissues were digested with Soluene 350 (Packard) and aliquots of the digestate counted in Dimilume (Packard) using an Intertechnique SL30 scintillation counter. We found that [^{14}C] HMPA was distributed throughout the whole body of rats following nasal instillation. The distribution profile is shown in Table 1 and is remarkably similar to that of methyl methanesulphonate (MMS). The doses of [^{14}C] HMPA used in these experiments were relatively high for experimental reasons, thus, while the data generated may be relevant to several of the effects seen in distant organs following intraperitoneal injection of HMPA, they do not preclude its selective accumulation in the nasal epithelium at the very low doses employed in the rodent carcinogenicity study ($\sim$0.02 ppm).

Direct (Hydrolytic) Formaldehyde Generators

Dinitrosopentamethylenetetramine (DNPT) This crystalline solid is formed from hexamethylenetetramine (HMT, see below) on treatment with nitrous acid and generates formaldehyde in aqueous media, probably via the formation of

Table 1   Distribution of Radioactivity after Nasal Administration of Either Hexamethylphosphoramide (HMPA) or Methyl Methanesulfonate (MMS)

Tissue	Total dose recovered (%)	
	HMPA	MMS
Liver	13.2	9.1
Blood	1.8	12.5
Nasal region	3.5	1.1
Kidneys	5.4	2.5
Lungs	0.8	2.2
Brain	0.5	0.4
Heart	0.2	0.7
Remaining carcass	74.6	71.5

Efficiency of counting was determined by the use of an internal standard. Results are expressed as a percentage of the total radioactivity recovered in the tissues.

a reactive methylol derivative (Table 2). It is noncarcinogenic in rats (9) by both oral administration (10-12) and intraperitoneal injection (13). Although DNPT is a nitrosamine, it should not be compared with DMN. It can act as a source of formaldehyde only, since the five methylene ($-CH_2-$) groups are restrained as an interrelated series of aminals, the collapse of which generates formaldehyde but no alkyldiazonium ions [*c.f.* DMN (4)]. DNPT gives a negative response in the *Salmonella* assay (6) and a positive response in the BHK cell transformation assay (Fig. 3*c*, ±S9).

Table 2   Release of Formaldehyde from Selected Compounds Expressed as a Percentage of Available $-CH_3$ or $-CH_2-$ Groups

Compound (µmoles)		Available $-CH_3$ or $-CH_2-$ released as HCHO (%)	
		+S9	−S9
HMPA(6)[*]	1	1.72	
	10	0.25	
DMN(2)	10	0.16	
	100	0.10	
DNPT(5)	0.1	86.8	103.8
	1.0	69.5	88.6
HMT(6)	0.1	73.6	81.8
	1.0	54.7	54.3

The compounds were incubated with S9 (*c.f.* Fig. 4) or buffer for 20 min at 37°C and the formaldehyde generated monitored by the Nash method (1).

[*]Number of available groups is shown in parentheses.

Hexamethylenetetramine (HMT) This solid is formed via condensation of formaldehyde with ammonia and it decomposes to these constituents in aqueous media (Table 2). Three oral carcinogenicity bioassays have been conducted with HMT, two by dietary administration (14, 15) and one via the drinking water (15); in each case a negative result was recorded. The contrasting positive result of a subcutaneous injection study with this material (16) will be discussed later in this article. HMT gives a positive response in the BHK cell transformation assay (Fig. 3d, ±S9).

Dimorpholinylmethane (DMM) Although the relevant studies have not been conducted, this material may be regarded as a direct generator of formaldehyde, similar to DNPT and HMT. Consistent with this suggestion, it gives a negative response in the *Salmonella* assay [R Trueman, unpublished data (1980)]. Preussman has reported that DMM produces local tumors following its subcutaneous injection in rats (17).

DISCUSSION

The S9-mediated release of formaldehyde from DMN and HMPA are compared in Fig. 4, and it is evident that HMPA generates formaldehyde more rapidly than DMN under the conditions of the experiment. Nonetheless, the release is much slower than the S9-independent release of formaldehyde by hydrolysis from either DNPT or HMT (Table 2). In addition, the equally high levels of formaldehyde released from both DNPT and HMT in the presence of S9 mix suggest that the low levels released from HMPA and DMN are not a direct consequence of further metabolism to formate ion or of rapid loss by reaction with the proteins contained within the S9 mix.

It was suggested earlier that the carcinogenicity of HMPA might be induced by the formaldehyde liberated during its metabolism and that the positive response observed for DNPT in the BHK cell transformation assay almost certainly was. Thus, comparison of a range of test responses observed for these two chemicals with those recorded for formaldehyde enables these possibilities to be pursued further. The data considered (6) are shown in Table 3 and are made particularly intriguing by the fact that DNPT, a prolific source of formaldehyde, appears to be noncarcinogenic, while HMPA, a seemingly minor source of formaldehyde, is a potent carcinogen.

The most obvious trend in Table 3 is that the in vitro activities of DNPT, HMPA, and formaldehyde are very similar. These similarities extend into the simplest of the in vivo assays available, i.e., the induction of mutations in *Drosophila*. In mammals, however, the activities of DNPT and HMPA diverge, with only the latter showing activity. An immediate explanation of this divergence is that DNPT has the potential to exert only local effects due to the rapidity with which the formaldehyde is released and either enzymatically or chemically

Table 3 Comparative Activities of Formaldehyde (I), Dinitrosopentamethylene-tetramine (DNPT, IV) and Hexamethylphosphoramide (HMPA, III) in a Range of In Vitro and In Vivo Bioassays

Point of reference	HCHO (I)	DNPT (IV)	HMPA (III)
Physical form	Gas $(bp_{760} - 19.5°C)$	Solid $(mp\ 213°C)$	Liquid $(bp_{11}\ 107°C)$
Decomposition to HCHO in vitro	100%	Rapid $(-S9)$	Slow $(+S9)$
In vitro assay responses			
Bacterial mutation assays	—	—	—
Eukaryotic assays			
Yeast (D7)	?	+	+
Yeast (JD1)	?	+	+
Yeast (RAD)	?	+	+
Mamm. cell mutation			
L5178Y/P388	+	+	+
Cytogenics (SCE)	?	+	+
BHK transformation	+	+	+
In vivo assay responses			
Drosophila melanogaster	+	+	+
Rodent assays			
Micronucleus assay (ip)	?	—	+
Micronucleus assay (ip)	?	—	+
Micronucleus assay (ip)	?	—	+
Carcinogenicity assay	+ (nasal)	—	+ (nasal)

These data are discussed in the text and are, in the main, abstracted from material presented in reference 6. The 3 micronucleus arrays were conducted independently.

destroyed. This is consistent with the observation that when DNPT is administered either orally or parenterally to rodents it has no effect on distant sites such as the bone marrow. Furthermore, the absence of even oral or peritoneal tumors, respectively, in these studies probably reflects the rapidity with which any formaldehyde generated is "inactivated" in these two environs. In contrast, HMPA is rapidly dispersed through many organs and tissues, in some of which it might slowly release formaldehyde, and this chemical shows cytogenetic activity in the bone marrow after its intraperitoneal injection into mice (6).

The possibility that the genetic toxicity of formaldehyde and the labile formaldehyde generators are expressed only in areas local to the point of administration is enhanced by the findings of two subcutaneous carcinogenicity studies reported for HMT (16) and DMM (17). This route of administration does not yield unambiguous evidence of the carcinogenicity of a chemical, especially in cases such as the present ones, in which local irritation would occur; nonetheless, the authors reported HMT and DMM to be potent subcutaneous carcinogens. This conclusion must remain tentative, but it reflects on the apparent non-carcinogenicity of DNPT (Table 2): if the positive in vitro responses seen for this chemical are considered evidence for its potential carcinogenicity, then the most sensitive test for carcinogenicity would be a carefully controlled subcutaneous injection study. However, the evaluation of such data in terms of human hazard assessment would be difficult.

The associations discussed above must remain tentative because several data points for the chemicals considered are either missing or inconsistent with other data. In particular three areas of uncertainty must be resolved before compounds III–VI (Fig. 1) can be accepted simply as alternative "formulations" of formaldehyde gas:

1 Cyclic analogues of HMPA. The carcinogenicity, mutagenicity, and cell-transforming properties of DMN are reflected in cyclic analogues such as N-nitrosomorpholine (Fig. 5). Likewise, earlier studies have shown that cyclic analogues of HMPA, such as tripiperidinylphosphine oxide, share the cell-transforming properties of HMPA in vitro (8) (Fig. 6). Clearly, the activity of such cyclic analogues of HMPA cannot be mediated by formaldehyde. This being so, it will be necessary either to expand consideration of the toxicology of formaldehyde to include complex aldehydes and methylol derivatives such as those that may form following enzyme-mediated α-hydroxylation of these cyclic analogues or to seek an alternative mechanism of carcinogenic action for HMPA.

2 The influence of S9 on the biological activity of DNPT and HMPA in vitro. Reference to Table 2 indicates that the quantity of formaldehyde formed

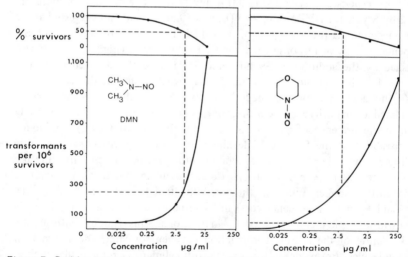

Figure 5 Positive test responses observed for dimethylnitrosamine (DMN) and N-nitrosomorpholine in the BHK cell transformation assay (+S9) (2, 8) (for experimental details see also legend to Fig. 3).

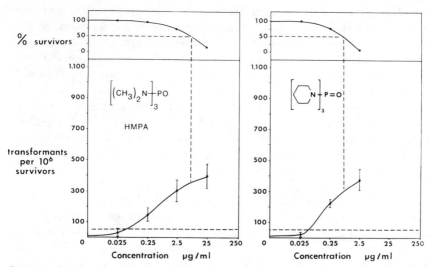

Figure 6 Positive test responses observed for hexamethylphosphoramide (HMPA) (7) and tripiperidinylphosphine oxide (8) in the BHK cell transformation assay (+ S9) (2, 8) (for experimental details see also legend to Fig. 3).

from DNPT or HMT is not greatly reduced by the addition of S9 mix. This was suggested earlier to indicate that the low yield of formaldehyde formed by the action of S9 mix on either DMN or HMPA was not due to its rapid destruction by the S9 itself. It is therefore interesting to compare the several BHK cell transformation responses observed for HMPA and DNPT. The positive response observed by Styles for HMPA (Fig. 3b) (8) was S9-dependent and suggested above to be mediated via the low levels of formaldehyde formed. In contrast the S9-independent positive response seen for DNPT by Daniel (Fig. 3c) was almost abolished by the addition of S9 mix, and Styles recently reported a negative response for this chemical in the same assay when tested in the presence of S9 (6) (no determination was made in its absence). Further, Styles also found the related formaldehyde generator DMM (VI) to be inactive when routinely tested in the presence of S9 mix [Styles, unpublished data (1980)]. Although it is possible that the formaldehyde liberated in these in vitro assays is more rapidly destroyed than it is when generated in the presence of semicarbazine as a trapping agent (1), these observations are at present inconsistent and therefore require further studies. Likewise, the possibility that the inactivity of formaldehyde (and formaldehyde generators) in bacterial mutation assays may be associated with its rapid metabolism to CO_2 in these organisms requires confirmation.

3 Nature of the chemical interaction between formaldehyde and cellular macromolecules. Formaldehyde can react with cellular macromolecules such as nuclear DNA or proteins in at least three separate ways (3). The first is to produce an electrophilic methylol adduct ($R-CH_2 OH$) (18), which can react with a

further molecule of substrate to produce a methylene-bridged adduct (R—CH$_2$—R). The formation from formaldehyde, or from a derived methylol derivative, of both DNA-DNA and DNA-protein bridges is consistent with both of these reactivities and is discussed further in Chapter 1. In addition to these direct reactions there is evidence for the biological oxidation of formaldehyde to dihydroxydimethylperoxide, a species capable of generating free radical peroxides that might lead to DNA damage of a different nature (19, 20). Auerbach (3) has cautioned that these separate reactivities of formaldehyde may become apparent under different conditions of test and, consequently, that positive test responses observed in different assays may be mechanistically unrelated. This possibility should be appreciated when extrapolating experimental data for formaldehyde from one test system to another, or to humans.

CONCLUSIONS

1 Formaldehyde, or a derived methylol or peroxide derivative, is capable of reacting with cellular macromolecules, is mutagenic to mammalian cells in vitro, and is carcinogenic at the site of contact (the nasal epithelium in the case of its inhalation).

2 Compounds that generate formaldehyde, either hydrolytically or enzymatically, possess some of the expected toxic properties of formaldehyde, especially when assayed in vitro.

3 The absence of overt carcinogenicity and some other forms of toxicity when hydrolytic formaldehyde generators are administered to rodents suggests that their genetic toxicity, and perhaps that of formaldehyde itself, is limited to the immediate area of initial tissue contact.

4 The possibility that the carcinogenicity of hexamethylphosphoramide (HMPA) is mediated by enzymatically released formaldehyde is worthy of further study. If this mechanism were to be confirmed, administration of HMPA would represent a unique method of delivering formaldehyde to body tissues distant from the point of application. The extreme carcinogenic potency of HMPA [HMPA is active at 0.02 ppm (5) as opposed to the lower effective dose of ~6 ppm for HCHO] may be related to its ability to enter mammalian cells and generate formaldehyde within the cell.

It is suggested that the continued study of "formaldehyde generators" might increase the general value of the detailed studies on formaldehyde described elsewhere in this volume.

Note added in proof: Low dose-levels of HMPA have now been shown to accumulate in the rodent nasal epithelium (Rickard and Gillies, The Toxicologist, 2, Abs. 162, 1982) and the present studies have now been formally described, including data for DMM (Ashby and Lefevre, Carcinogenesis, *in press* 1982).

REFERENCES

1 Nash, T. The colorimetric estimation of formaldehyde by means of the Hantzsch reaction. *Biochem. J.* 55:416–421 (1953).

2 Styles, JA. A method for detecting carcinogenic organic chemicals using mammalian cells in culture. *Br. J. Cancer* 36:558–563 (1977).

3 Auerbach, C, M Moutschen-Dahmen, and J Moutchen. Genetic and cytogenetic effects of formaldehyde and related compounds. *Mutat. Res.* 39: 317–362 (1977).

4 Preussman, R, KN Arjungi, and G Ebers. In vitro detection of nitrosamines and other indirect alkylating agents by reaction with 3,4-dichlorothiophenol in the presence of rat liver microsomes. *Cancer Res.* 36:2459–2462 (1976).

5 Lee, KP, and Trochimowicz, HJ, Induction of nasal tumors in rats exposed to hexamethylphosphoramide by inhalation. *J. Natl. Cancer Inst.* 68:157–164 (1982).

6 *Short-Term Tests for Carcinogens: Report of the Internation Collaborative Program*, edited by FJ de Serres, and J Ashby. Amsterdam: Elsevier-North Holland (1980).

7 Ashby, J, JA Styles, and D Anderson. Selection of an in vitro carcinogenicity test for derivatives of the carcinogen hexamethylphosphoramide. *Br. J. Cancer* 36:564–565 (1977).

8 Ashby, J, JA Styles, and D Paton. Potentially carcinogenic analogues of the carcinogen hexamethylphosphoramide: Evaluation in vitro. *Br. J. Cancer* 38:418–427 (1978).

9 Dinitrosopentamethylenetetramine. *1 ARC Monographs on the Evaluation of Carcinogenic Risk of Chemicals to Man* 11:241–245 (1976).

10 Griswold, DP, Jr, AE Casey, EK Weisburger, JH Weisburger, and FM Schabel, Jr. On the carcinogenicity of a single intragastric dose of hydrocarbons, nitrosamines, aromatic amines, dyes, coumarins, and miscellaneous chemicals in female Sprague-Dawley rats. (1977).

11 Weisburger, JH, EK Weisburger, N Mantel, Z Hadidian, and TN Frederickson. New carcinogenic nitrosamines. *Naturwissenschaften* 53:508 (1966).

12 Hadidian, Z, TN Fredrickson, EK Weisburger, JH Weisburger, RM Glass, and N Mantel. Tests for chemical carcinogens. Report on the activity of derivatives of aromatic amines, nitrosamines, quinolines, nitroalkanes, amides, epoxides, aziridines and purine antimetabolites. *J. Natl. Cancer Inst.* 41:985–1036 (1968).

13 Boyland, E, RL Carter, JW Gorrod, and FJC Roe. Carcinogenic properties of certain rubber additives. *Eur. J. Cancer* 4:233–239 (1968).

14 Natrig, H, J Andersen, and EW Rasmussen. A contribution to the toxicological evaluation of hexamethylenetetramine. *Food Cosmet. Toxicol.* 9:491–500 (1971).

15 Porta, GD, MI Colnaghi, and G Parmiani. Non-carcinogenicity of hexamethylenetetramine in mice and rats. *Food Cosmet. Toxicol.* 6:707–715 (1968).

16 Grasso, P, and L Golberg. Subcutaneous sarcoma as an index of carcinogenic potency. *Food Cosmet. Toxicol.* 4:297–320 (1966).

17 Preussman, R. Direct alkylating agents as carcinogens. *Food Cosmet. Toxicol.* 6:576–577 (1968).

18 Hendry, JA, FL Rose, and AL Walpole. Cytotoxic agents. I. Methylolamides with tumour inhibitory activity, and related inactive compounds. *Br. J. Pharmacol.* 6:201–234, 1951).

19 Jenson, KA, K Kirk, G Kolmark, and M Westergaard. *Cold Spring Harbor Quant. Biol.* 16:245 (1951).

20 Sobels, FH. Organic peroxides and mutagenic effects in Drosophila. *Nature* 117:979–980 (1956).

21 Anderson, D, and M Fox. The induction of thymidine and IUdR resistant mutants in P388 mouse lymphoma cells by X-rays, UV and mono- and bifunctional alkylating agents. *Mutat. Res.* 25:107–122 (1974).

Chapter 10

A 26-Week Inhalation Toxicity Study with Formaldehyde in the Monkey, Rat, and Hamster

George M. Rusch

Henry F. Bolte

William E. Rinehart

There have been numerous reports of possible long-term exposures to formaldehyde, particularly as a consequence of the use of urea-formaldehyde foam insulation and particleboard and plywood by the building trade (1). This study was designed to investigate the toxic effects of low levels of formaldehyde vapor on multiple species of animals when administered in a manner that would approximate an environmental exposure to the general public.

METHOD AND MATERIALS

Groups of 6 male cynomologus monkeys, 20 male and 20 female Fischer 344 rats, and 10 male and 10 female Syrian Golden hamsters were exposed to vapors of formaldehyde at target concentrations of 0, 0.20, 1.00, and 3.00 ppm,

Performed at Bio/dynamics, East Millstone, New Jersey under contract from The Formaldehyde Institute, Scarsdale, New York.

The authors wish to acknowledge the contributions toward study performance of Mr. Gary Hoffman, Mr. Philip Bini, Mr. Kim Nitahara, Mr. Douglas Reigle, Mr. Fred Whitman, Mr. Richard Stagg, Ms. Julia Davis, and Ms. Rose Buehler.

respectively 22 h/day, 7 days/week for 26 weeks. During the first 6 weeks of the study the initial high level group was terminated due to an uncertainty in the determination of the exposure concentration. This group was subsequently replaced with a new exposure group and a corresponding control group. The study has, therefore, been divided into two segments, with the 0.20 and 1.00 ppm exposure groups being compared to their control group and the 3.00 ppm exposure group being compared to its control group.

The test material was supplied as an unstabilized, aqueous, 5 percent solution of formaldehyde by one of the sponsoring companies. An analysis of this solution, by the supplier, gave average values of 4.94 percent formaldehyde, 0.03 percent methanol, 6 ppm formic acid, 120 ppm total ash, 9 ppm of chromium, 15 ppm of sodium, and less than 6 ppm of all other metals combined. The solution was assayed in our laboratory for formaldehyde daily during the first 9 days of the study and weekly thereafter until termination of the study. The mean concentration of formaldehyde was found to be 4.94 percent, with a standard deviation of 0.11 percent. No change in formaldehyde concentration was detected during the study.

The exposure concentrations of formaldehyde were generated using a double bubbler system (Fig. 1). The system consisted of two bubblers containing the 5 percent formaldehyde, each immersed in a separate constant temperature bath. Compressed air was passed through a flowmeter, then through a copper coil immersed in the first water bath, through the formaldehyde solution in the first bubbler, next through the formaldehyde solution in the second bubbler, and finally through a heated transfer line to the exposure chamber air intake line. The copper coil served to warm the air prior to it entering the first bubbler. This helped to maintain the temperature equilibrium in the bubbler. The second bubbler was maintained at a lower temperature than the first. Depending on the concentration of formaldehyde in the air stream, this bubbler could either contribute additional material or scrub out excess material. From there the heated transfer line served to prevent condensation of both water vapor and formaldehyde prior to induction into the main chamber air supply duct. Using

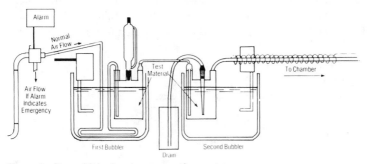

Figure 1 Formaldehyde vapor generation system.

Table 1 Generation Conditions

Group	Concentration (ppm)	Temperature (°C)	
		Bubbler 1	Bubbler 2
II	0.2	50	35
III	1.0	60	40
VI	3.0	70	50

this system it was felt that the formaldehyde concentration in the air stream would approach saturation, and thus the quantity delivered to the chamber would be directly proportional to the airflow through the bubblers. Typically, formaldehyde solution was added to the first bubbler and withdrawn from the second bubbler during the exposure. Attempts to assay the formaldehyde concentrations in each bubbler following the exposures, and thus calculate a nominal concentration, were unsuccessful, since the system was designed to consume only a small amount of the available formaldehyde and changes in formaldehyde concentration in the solution were small. The temperatures for each bubbler are given in Table 1. This 15-20 degree difference in temperature provided the most stable generating conditions.

The exposure concentrations were measured eight times per day using a modified NIOSH chromotropic acid method (2) for analysis of airborne levels of formaldehyde. The sample train used for collection is shown in Fig. 2. The two modifications of the NIOSH method consisted of using a 1 percent aqueous sodium bisulfite solution in the bubbler instead of water and of warming the reactants in a boiling water bath for 10 min instead of allowing the heat of mixing alone to push the reaction to completion. The sodium bisulfite solution was used since it both stabilized the formaldehyde and improved scrubbing efficiency (Meadows and Rusch, unpublished data). The boiling water bath provided more uniform reaction conditions and hence better reproducibility.

Samples were drawn through the bubblers at a rate of from 1.5 to 2.0 liters/min. The volume of sample was different for each chamber to allow for collection of approximately equal amounts of formaldehyde (Table 2). The actual formaldehyde concentration was calculated by comparing the adsorbance of the

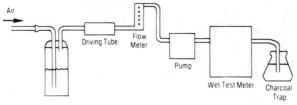

Figure 2 Formaldehyde vapor sample collection system.

Table 2   Chamber Atmosphere Sample Collection Conditions

Group	Target concentration (ppm)	Sample vol. (liter)	Scrubber vol. (ml)	Nominal CH_2O concentration ($\mu g/ml$)
I, V	0	60	10	
II	0.20	30	10	0.74
III	1.00	20	20	1.23
IV	3.00	10	30	1.23

solution of the final chromotropic acid reaction product to a calibration curve prepared using standard formaldehyde solutions (Fig. 3). This curve was linear over the range of from 0 to 3.5 μg formaldehyde per milliliter of solution. The concentration of formaldehyde in the exposure chambers was calculated using the equation shown below:

$$\text{ppm} = \frac{\text{Conc. } (\mu g/ml) \times \text{sample solution vol. (ml) } 24.5}{\text{sample air volume (liters) } 30.03}$$

where 24.5 = gas law constant at room temperature and 30.03 = molecular weight of formaldehyde. Both the concentration and sample solution volume shown in the equation above were for the aqueous sodium bisulfite used in the bubbler and do not include the volumes of the reagents used for the colorimetric analysis.

In addition to the eight grab samples collected from each chamber, per exposure, each chamber was monitored continuously using a CEA Model 555 monitor. The output from this instrument was simultaneously fed to a strip

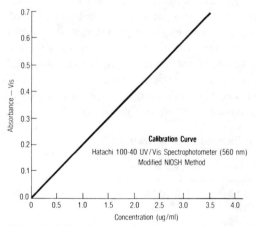

Figure 3   Calibration curve for chromotropic acid analysis of formaldehyde.

Table 3   Target Specifications for Exposure Concentrations

Target concentration	Individual samples-max.	Daily mean ranges	Alarm settings		
			Min.	Max.	Cutoff
0.20	0.30	0.17–0.23	0.16	0.24	0.26
1.00	1.50	0.80–1.20	0.8	1.2	1.3
3.00	3.90	2.70–3.30	2.6	3.4	3.6

chart recorder and an alarm system. The alarm system was designed to provide an audible signal in the event of a shift in concentration in either direction and to divert the airflow through the bubblers in the event of a significant upward shift in concentration. The alarm settings and target specifications are shown in Table 3. The low level chamber exceeded the 0.30 maximum 17 times in 1456 individual samples, the mid level chamber exceeded the 1.5 maximum four times, and the high level chamber exceeded the 3.9 ppm maximum 13 times. While the daily mean concentration for the low level chamber was outside target 21 times in 182 exposure days, the mid level chamber was outside target only twice and the high level chamber was outside target only 17 times. None of these excursions was considered to have had an effect on the study. In the control groups the average instrument response measured as formaldehyde was 9 parts per billion. In the room, the average response measured as formaldehyde was 17 ppb. Both values were considered to represent normal ambient background levels of formaldehyde combined with other substances capable of reacting with the chromotropic acid reagent. In the low level exposure group (0.20 ppm) the cumulative mean concentration was 0.19 ppm. In the mid level exposure group (1.00 ppm) the cumulative mean concentration was 0.98 ppm. In the high level exposure group (3.00 ppm) the cumulative mean concentration was 2.95 ppm.

In addition to samples analyzed in our laboratory, approximately 65 samples from each of the low and mid level exposure groups and 103 samples from the high level group were split. One-half of each sample was analyzed in our laboratory and one-half sent to one of the sponsoring companies' analytical laboratories for independent analysis. The average results of a comparison of these analyses along with the average of the individual values obtained from our CEA monitor for the same time intervals are shown in Table 4.

Table 4   Comparative Analysis of Exposure Concentrations

Group	Bio/dynamics	Sponsor	CEA monitor
II	0.18	0.19	0.19
III	0.91	0.89	0.97
VI	2.96	2.78	2.90

Table 5 Mortality

Group	Monkey	Rat	Hamster
I	0	0	1
II	0	0	2
III	0	0	2
V	0	1	1
VI	0	0	1

RESULTS

Seven hamsters and one rat died during the study (Table 5). These deaths were scattered through all exposure and control groups and were not treatment related. The deaths all occurred during the last 6 weeks of the study, and while a definite cause was not established for all animals, in a few cases death appeared to be due to renal failure.

All animals were observed three times daily and given a full written assessment once per week. A summary of the most significant findings for the monkeys is given in Table 6. The animals in the high dose group showed an increased incidence of hoarseness, congestion, and nasal discharge. Some monkeys in the mid and low level groups also showed increased incidences of nasal discharge. These latter findings were limited to from one to two animals per group, generally were observed during the middle of the study, and do not appear to represent treatment-related effects. In contrast, the findings in the high level group appear to be related to the formaldehyde exposures and while generally observed throughout the study, they tended to be more numerous during the last 13 weeks of exposure.

Observations of the rats, summarized in Table 7, did not show evidence of treatment-releated effects. There was a high incidence of chromodacryorrhea and lacrimation, especially in the first control group. Both the chromodacryorrhea and lacrimation were noted more frequently in the female rats than in the

Table 6 Summary of Physical Observations in Monkeys

Observation	Total incidence by group[*]				
	I	II	III	V	VI
No observed abnormalities	139	98	93	138	64
Hoarseness	0	0	0	0	32
Congestion	0	0	0	0	36
Nasal discharge	9	30	45	5	62

[*]Out of a total of 156 observations per group.

Table 7 Summary of Physical Observations in Rats

Observation	Total incidence by group[*]				
	I	II	III	V	VI
No observed abnormalities	560	663	754	655	614
Rales	24	39	35	50	65
Nasal discharge	76	103	89	50	55
Chromodacryorrhea	182	96	28	74	38
Lacrimation	187	119	62	99	37

[*]Out of a total of 1040 observations per group.

male rats; also, while reported throughout the study, the incidence was greater in the last half of the study.

Observations of the hamsters (Table 8) showed slightly higher frequency of rales, nasal discharge, and lacrimation in the formaldehyde-exposed animals than was observed in the corresponding control groups. However, the increase in frequency was slight when compared to the total number of 540 observations.

Body weights of the monkeys (Table 9) in all groups were normal throughout the study. The body weights of the male rats (Table 10) in the high level group were depressed to a statistically significant degree from the second week of the study through termination when compared to the respective control group. By the end of the study, this difference was almost 20 percent and is considered to be related to the formaldehyde exposures. It is also interesting to note that the initial body weights of the second control group and 3.0 ppm exposure group (groups V and VI, respectively) were approximately 20 g less than the initial body weights for the first control group, the 0.20 ppm exposure group, and the 1.0 ppm exposure group (groups I, II, and III, respectively). While this difference appears to represent about 10 days' growth, there was only a 2-day difference in the age of the animals when they were put on test, i.e., 48 days versus 50 days. Similar differences were observed in the initial female rat body weights and in the hamster body weights that follow. It is doubtful that

Table 8 Summary of Physical Observations in Hamsters

Observation	Total incidence by group[*]				
	I	II	III	V	VI
No observed abnormalities	458	444	399	358	376
Rales	1	7	11	5	23
Nasal discharge	6	18	23	7	13
Lacrimation	1	4	12	3	4

[*]Out of a total of 520 observations per group.

Table 9 Representative Mean Body Weights of Monkeys

Group	Pre	Wk 1	Wk 2	Wk 8	Wk 14	Wk 20	Wk 26
I	2.5	2.4	2.5	2.7	2.8	3.0	3.2
II	2.4	2.2	2.4	2.6	2.8	2.9	3.0
III	2.4	2.3	2.5	2.7	2.8	3.0	3.2
V	2.5	2.5	2.6	2.7	2.9	3.0	3.2
VI	2.4	2.6	2.6	2.7	2.9	3.1	3.2

the cause could be related to a problem with the supplier, since the rats were purchased from Charles River in Wilmington, Massachusetts and the hamsters from Charles River Lakeview in Newfield, New Jersey. It is possible that this may be related to the fact that the rats in the earlier shipment were born in the spring (on May 1, 1979) and the later group were born in the summer (on August 14, 1979), while the respective birth dates for the hamsters were May 8, 1979 and August 21, 1979; thus we may have seen a seasonal variation in body weight. The body weights of the female rats (Table 11) were similar to those of the male rats, with the high exposure group showing statistically significant lower body weights of about 10 percent starting in the second week of the study and continuing through the termination of the study.

With the exception of the differences in starting weights discussed above, the male (Table 12) and female (Table 13) hamster body weights were unremarkable.

Absolute and relative adrenal, heart, kidney, and liver weights were compared for all species. In addition, some lung weights from groups I, II, and III animals were measured. Lung weights were not obtained from animals whose lungs were prepared for electron microscopic examination, since these lungs required very rapid fixation. All organ weights were considered to be unremarkable, except for the liver weights from the 3.0-ppm rats (Table 14). In the 3-ppm group, both absolute and relative liver weights were lower than the corresponding control values. In the male rats, the liver weights were approximately 26 percent lower and in the female rats the difference was approximately 12

Table 10 Representative Mean Body Weights of Male Rats

Group	Pre	Wk 1	Wk 2	Wk 3	Wk 8	Wk !$	Wk 20	Wk 26
I	131	170	199	223	298	337	361	388
II	134	171	202	224	301	345	365	400
III	134	168	192	215	289	320	356	383
V	112	121	152	181	271	329	360	397
VI	113	114	138*	163*	232*	281*	313*	332

*Significantly different from control.

Table 11   Representative Mean Body Weights of Female Rats

Group	Pre	Wk 1	Wk 2	Wk 3	Wk 8	Wk 14	Wk 20	Wk 26
I	109	126	139	148	182	198	206	215
II	107	125	139	151	187	204	216	227
III	107	123	134*	144	176	186	204	215
V	92	98	116	129	169	193	206	217
VI	94	95	107*	119*	156*	177*	183*	198*

*Significantly different from control.

Table 12   Representative Mean Body Weights of Male Hamsters

Group	Pre	Wk 1	Wk 2	Wk 8	Wk 14	Wk 20	Wk 26
I	79	89	95	126	140	148	149
II	77	84	91	128	144	154	148
III	73	84	91	124	140	148	150
V	61	65	75	109	126	133	135
VI	62	65	75	115	131	138	141

Table 13   Representative Mean Body Weights of Female Hamsters

Group	Pre	Wk 1	Wk 2	Wk 8	Wk 14	Wk 20	Wk 26
I	79	87	93	127	141	150	148
II	78	90	95	134	150	153	150
III	76	85	92	133	153	155	161
V	65	67	79	114	132	142	141
VI	65	69	78	114	133	141	147

Table 14   Comparison of Absolute and Relative Liver
Weights in Rats

	Males		Females	
Group	Wt. (g)	Ratio (%)	Wt. (g)	Ratio (%)
I	12.3	3.16	7.19	3.33
II	12.5	3.11	7.55	3.32
III	11.7	3.04	7.20	3.36
V	12.5	3.15	7.29	3.30
VI	9.9*	2.90*	6.51*	3.17

*Significantly different from control.

percent. The difference in the ratios was not as great since the terminal body weights in this group were also depressed. Decreased heart and kidney weights were also observed in this group; however, since the relative weights for these organs were increased, these findings were attributed to the depressed terminal body weights.

Gross necropsy examinations were performed on all animals. The organs most frequently reported with abnormalities were the lungs, liver, and kidneys. Generally, these findings consisted of discolorations or the presence of multiple foci. They were scattered through all groups and were of the type typically seen in experimental animals. None appeared to indicate a response to the exposures.

Microscopic examinations were performed on the nasal turbinates and all gross lesions in all animals. The lungs and trachea were also examined in the animals in groups I, III, V, and VI. Microscopic examination of all tissues, except nasal turbinates, discussed below, showed lesions that have frequently been observed in laboratory animals and were not considered to be related to the exposures. This included the rat liver, although only livers reported with gross abnormalities were examined. This included only 8 of the 40 rats in the 3.0-ppm group.

Transmission electron microscopic examinations were performed on sections of the left diaphragmatic and right apical lobes of the lung, the trachea and epithelium of the right side of the nasal turbinate from five rats per sex from groups I and III. None of the sections examined revealed evidence of treatment-related effects.

Initially, one section of the nasal turbinate was examined by light microscopy from each animal in groups I, III, V, and VI. When the results from these examinations indicated a possible response in the 3.0-ppm rats, additional sections were examined from all animals. Specifically, as shown in the sagittal section of the rat turbinate (Fig. 4), three transverse sections were taken from the nasoturbinate, one each from the anterior (A), middle (B), and posterior (C) regions and one from the ethmoturbinates (E). When possible, sections from similar regions of the turbinate of the hamsters and monkeys were also examined. In the hamsters these findings have also been evaluated by region; in the monkey, combined findings from regions A, B, and C are presented for each animal.

Examinations of sections from the turbinate of the rat (Table 15) showed

Figure 4 Sagittal section through the nasal and paranasal cavities of the rat.

Table 15   Significant Findings in Nasal Turbinates of Rats

Group	Level	Squamous metaplasia				Basal cell hyperplasia			
		A	B	C	E	A	B	C	E
I	0	19/38	2/38	0/33	0/37	4/38	0/38	0/33	0/37
II	0.20	1/39	1/38	0/38	0/37	0/39	0/38	1/38	0/37
III	1.00	12/38	3/36	0/32	0/34	4/38	0/36	0/32	0/34
V	0	7/40	3/39	0/38	0/38	2/40	4/39	0/38	0/38
VI	3.00	31/39	23/37	3/37	0/38	9/39	25/37	2/37	0/38

frequent incidences of squamous metaplasia and squamous hyperplasia (both classified as squamous metaplasia in Table 15) and of basal cell hyperplasia. A high incidence of squamous metaplasia was observed in the A sections from the anterior nasoturbinate in most groups. A comparison of these incidences in the control, 0.20-, and 1.0-ppm groups did not appear to show a treatment-related response, while a comparison of the incidences between the control and 3.0-ppm group showed a fourfold increase in the exposed group. It should be remembered that stratified squamous epithelium is a normal component of this region of the nasal mucosa, and, therefore, this is not the best area to examine for squamous metaplasia. Additionally, the low incidence in the 0.20-ppm group may be attributable to the fact that these turbinates had not originally been sectioned for the single section evaluation, and in sectioning them, the A section may have been taken farther from the vestibular area than sections in the other groups. The incidence of basal cell hyperplasia in this section did not show evidence of treatment-related effects in the 0.20- and 1.0-ppm groups but showed approximately a fourfold increase for the 3.0-ppm group compared to its control. The only group that showed evidence of squamous metaplasia in the C sections, the posterior nasoturbinate, was the 3.0-ppm group. While the incidence was low, this would appear to support the observations in the other areas of the nasoturbinate. Findings in the ethmoturbinate were unremarkable. The comparison of sections taken from the middle region of the nasoturbinate,

Table 16   Rhinitis in Rats

Group	Level	Section			
		A	B	C	E
I	0	17/38	5/38	1/33	1/37
II	0.20	14/39	7/38	5/38	0/37
III	1.00	14/38	6/36	3/32	0/34
V	0	12/40	14/39	1/38	1/38
VI	3.00	25/39	19/37	2/37	0/38

Table 17 Significant Findings in Nasal Turbinates of Hamsters

Group	Level	Squamous metaplasia				Basal cell hyperplasia			
		A	B	C	E	A	B	C	E
I	0	0/14	1/19	0/17	0/17	0/14	0/19	0/17	0/17
II	0.20	0/4	1/14	0/17	0/18	0/4	1/14	0/17	0/18
III	1.00	0/11	2/19	0/15	0/17	0/11	2/19	0/15	0/17
V	0	0/9	0/20	1/16	0/18	0/9	0/20	0/16	0/18
VI	3.00	1/16	2/17	0/15	0/17	1/16	0/17	0/15	0/19

section B, most clearly defined the response. The high incidence of squamous metaplasia in the 3.0-ppm group (23 of 37), an eightfold increase compared to its control group (3 of 39), strongly indicated a response to the exposures. These findings were confirmed by also looking at the incidence of basal cell hyperplasia in the same two groups. Again there was a marked increase in the 3.0-ppm group (25 of 37), when compared to its control group (4 of 39). In most cases the squamous metaplasia and basal cell hyperplasia were judged to be mild to moderate, although they appeared to be somewhat more severe in the 3.0-ppm group.

Since the test material, formaldehyde, is a highly water-soluble irritant gas, comparisons were made of the incidence of rhinitis, as characterized by the presence of variable numbers of neutrophils and lymphoid cells (Table 16). Again the only treatment group showing an increased evidence, when compared to the corresponding control group, was the high level exposure group, primarily in the anterior and middle sections of the nasoturbinate.

Comparison of the incidence of squamous metaplasia (Table 17) in the hamster did now show evidence of exposure-related effects, although there were a few scattered observations of metaplasia in all groups. Likewise the incidence of rhinitis (Table 18) did not indicate a marked response to the exposures.

Evaluation of sections from the turbinates of the monkeys (Table 19) showed evidence of a response to the exposure only in the high level exposure

Table 18 Rhinitis in Hamsters

Group	Level	Section			
		A	B	C	E
I	0	0/14	1/19	1/17	0/17
II	0.20	0/4	0/14	0/17	0/18
III	1.00	0/11	2/19	0/15	0/17
V	0	0/9	0/20	0/16	0/18
VI	3.00	2/16	1/17	0/15	0/19

Table 19 Significant Findings in Nasal Turbinates of
Monkeys

Group	Level	Squamous metaplasia	Rhinitis
I	0	0/6	4/6
II	0.20	0/6	4/6
III	1.00	1/6	5/6
V	0	0/6	2/6
VI	3.00	6/6	4/6

group, characterized by squamous metaplasia and hyperplasia. Incidence of rhinitis in these animals was scattered through all groups and did not appear to be treatment related.

SUMMARY

In summary, groups of monkeys, rats, and hamsters were exposed to formaldehyde vapors at target concentrations of 0.2, 1.0, and 3.0 ppm for 22 h/day, 7 days/week for 26 weeks. The cumulative mean concentrations were 0.19, 0.98, and 2.95 ppm, respectively. In the monkeys the most significant findings were observations of squamous metaplasia in the 3.0-ppm exposure group. Observations of hoarseness, congestion, and nasal discharge were also seen in these animals, indicating irritation. In the rats the most significant findings were also increased incidences of squamous metaplasia and basal cell hyperplasia of the mid region of the nasoturbinate in the 3.0-ppm group. In addition, marked body weight gain depressions were observed again in the 3.0-ppm group, starting during the second week of the study and continuing through termination. No treatment-related effects were observed in the 0.2- and 1.0-ppm groups. In contrast to the monkeys and the rats, the hamsters did not show any significant responses to the exposures, even at the 3.0-ppm exposure level.

REFERENCES

1 Committee on Toxicology for Consumer Product Safety Commission. *Formaldehyde—An Assessment of Its Health Effects*. National Academy of Sciences (March 1980).
2 *Manual of Analytical Methods*. Health, Education and Welfare Publication No. (NIOSH) 75-121, 125 (1974).

Chapter 11

The Chronic Effects
of Formaldehyde Inhalation
in Rats and Mice: A Preliminary
Report

William D. Kerns

David J. Donofrio

Kenneth L. Pavkov

Formaldehyde is the most important commercially produced aldehyde in the United States, with over 9 billion pounds produced annually (1). Because of the common uses of formaldehyde in building materials, textiles, insulation, and several other industries, there is potential for occupational and environmental exposure (2). Considerable human exposure to formaldehyde vapor occurs at concentrations around 1 ppm (3). Formaldehyde is known to cause eye, nose, and throat irritation (4) as well as dermal irritation and allergic contact dermatitis (5, 6).

Adequate toxicity and carcinogenicity studies in animals have not been reported. When Sprague-Dawley rats were exposed concurrently to formaldehyde (14.7 ppm) and hydrochloric acid (10.7 ppm), nasal carcinomas developed in 25 percent of the animals (7).

Formaldehyde is mutagenic in some bacteria, fungi, and *Drosophila* (8), and

Supported by The Chemical Industry Institute of Toxicology, Research Triangle Park, North Carolina 27709.

The authors wish to acknowledge the following persons who participated in one or more phases of this study: Daryl Thake, Gerald Fisher, Ron Joiner, Ralph Mitchell, Melanie Connell, Grace Henry, and Susan Icely.

it induces unscheduled DNA synthesis in HeLa cells (9). Formaldehyde also appears to induce DNA-protein cross-linkages, which are removed by DNA repair enzymes in *Escherichia coli* repair-proficient strains (10). Formaldehyde was not mutagenic in the Chinese hamster ovary (CHO) assay (11) but did induce sister chromatid exchange in cultured CHO cells and human lymphocytes (12).

Long-term toxicity and carcinogenicity inhalation studies in rats and mice were initiated in order to more accurately assess the chronic effects of formaldehyde exposure. This discussion presents the most current findings in rats and mice based on data available after 27 months of investigation.

MATERIALS AND METHODS

Animals and Exposure

Seven-week-old Fischer 344 rats (Charles River Laboratories, Portage, MI) and 6-week-old B6C3F1 mice (Charles River Laboratories, Wilmington, MA) were randomly assigned to three exposure groups (target concentrations of 2, 6, and 15 ppm) and one control group (0 ppm). There were 119–121 animals of each sex in each of the exposure and control groups.

All animals were exposed for 6 h/day, 5 days/week, for a period of up to 24 months. This exposure period was followed by a 6-month period of non-exposure. Scheduled interim sacrifices were conducted at 6, 12, 18, 24, and 27 months, and the study was terminated at 30 months (Tables 1 and 2). The exposures took place in 5 cubic meter Hinners-type chambers that were operated at approximately 1 inch H_2O negative pressure with 12 air changes per hour. The formaldehyde concentrations, generated by heating paraformaldehyde (Aldrich Chemical Co., Inc., Milwaukee, WI), were monitored with a Miran 1-A infrared spectrophotometric gas analyzer. The internal environment was maintained at $20°-22°C$ and $51 \pm 5\%$ humidity. The cage positions were rotated

Table 1   Numbers of Rats That Were Killed at Scheduled Intervals

| Time interval | Exposure group and sex | | | | | | | |
| | 0 ppm | | 2 ppm | | 6 ppm | | 15 ppm | |
	M	F	M	F	M	F	M	F
6 Mo	10	10	10	10	10	10	10	10
12 Mo	10	10	10	10	10	10	10	10
18 Mo	20	20	20	20	20	20	20	19
24 Mo	54	47	50	44	41	41	13	14
27 Mo	10	9	10	10	10	9	5	0
30 Mo	6	4	4	1	3	5	0	0

Table 2   Numbers of Mice That Were Killed at Scheduled Intervals

Time interval	Exposure group and sex							
	0 ppm		2 ppm		6 ppm		15 ppm	
	M	F	M	F	M	F	M	F
6 Mo	10	10	10	10	10	10	10	10
12 Mo	10	10	10	10	10	10	10	10
18 Mo	0	20	1	20	0	20	0	19
24 Mo	21	31	22	26	20	41	17	28
27 Mo	0	16	0	12	0	11	0	9
30 Mo	0	0	0	0	0	0	0	0

one position from top to bottom and left to right each day throughout the exposure period. The actual concentrations achieved over 24 months were 14.3, 5.6, 2.0, and 0 ppm, respectively (Table 3).

Rats were housed individually in stainless-steel wire mesh cages and mice were similarly housed with four animals of the same sex per cage. The rats and mice were fed Purina Rodent Chow 5001 and allowed free access to water during the nonexposure hours.

All animals were observed twice daily throughout the study, and body weight determinations were made weekly for the first 6 months and biweekly thereafter. The rats were weighed individually and the mice were weighed by compartment groups.

Ophthalmoscopic examinations, hematology, serum chemistry and urinalysis determinations, and neurological function examinations were also made at selected intervals in the study. The results of these data will be presented in other related publications.

Pathology

Gross pathological examinations were performed on all animals that died (Table 4) or were killed at scheduled intervals (Tables 1 and 2) during the course of the

Table 3   Mean Chamber Concentration of Formaldehyde Vapor in Rats and Mice

Target (ppm)	Mean (ppm)	No.	SD	SE	Maximum	95% confidence interval
15	14.3	5811	2.77	0.04	32.4	14.2 to 14.3
6	5.6	5842	1.22	0.02	13.4	5.6 to 5.7
2	2.0	5838	0.57	0.01	6.2	2.0 to 2.1

Table 4   Summary of Unscheduled Deaths That Occurred
During the First 24 Months of the Formaldehyde Study

Exposure group	Rats		Mice	
	M	F	M	F
0 ppm	6	13	78	30
2 ppm	10	16	77	34
6 ppm	19	19	81	19
15 ppm	57	67	82	34

Note: M = male; F = female.

study. All major tissues from each organ system (approximately 40 per animal) in the 0-ppm and 14.3-ppm exposure groups were evaluated histologically. The tissues were fixed in 10 percent neutral buffered Formalin, embedded in paraffin, sectioned to a thickness of 5 μm, and stained with hematoxylin-eosin. Multiple sections of nasal turbinates were evaluated as target tissues in all rats and mice. Tissue masses that were observed at necropsy were evaluated microscopically in all animals.

Evaluation of Nasal Turbinates

The nasal turbinates from both rats and mice were processed in a similar manner. Histological sections were evaluated from five anatomical levels in the rat and from three levels in the mouse. Before immersion of the head in 10 percent neutral buffered Formalin, the nasal cavity was flushed with Formalin in a retrograde fashion through the nasopharyngeal orifice. After fixation for at least 5 days, the heads were decalcified (Decal, Omega Scientific Corporation, New Rochelle, NY), and sectioned. The rats required 48 h of decalcification and the mice required 24 h before further processing.

To insure that similar sections were evaluated from consecutive animals, the trimming procedure for the nasal cavity was standardized so that nearly identical sections were taken from each animal. After decalcification, the nasal cavity was trimmed at the appropriate levels indicated in Figs. 1 and 2. In the rat, level I was positioned in the cassette with the posterior surface down and the remaining sections were positioned with their anterior surfaces down. Levels II, III, and V were routinely processed for the mice. The cassettes were submitted to the histology laboratory for routine processing.

RESULTS

Rats

The 5.6-ppm and 14.3-ppm exposure groups demonstrated an apparent exposure-related decrease in body weight over the 2-year period in males and females

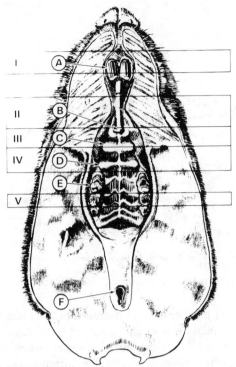

A = INCISOR TEETH   D = SECOND PALATAL RIDGE
B = INCISIVE PAPILLAE   E = FIRST MOLAR
C = FIRST PALATAL RIDGE   F = NASOPHARYNGEAL
                              ORIFICE

Figure 1 Ventral view of the hard palate from the rat demonstrates the appropriate levels to be processed for histology.

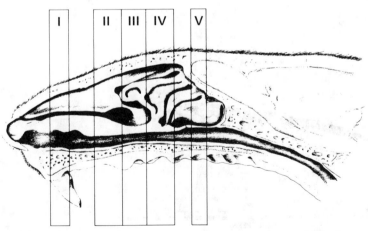

Figure 2 Midsagittal section of a rat head demonstrates the turbinates that are included in each level for microscopic evaluation.

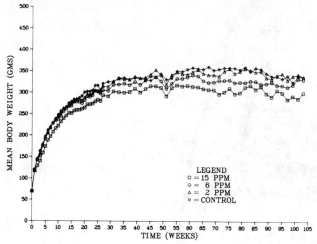

Figure 3 Mean body weight for male rats exposed to formaldehyde vapor.

(Figs. 3 and 4). At the 12-month interim sacrifice decreases in body weights of all groups were associated with typical histomorphological lesions of sialodacryoadenitis virus infection (Figs. 3 and 4). In the surviving animals, these weight decreases were temporary and there apparently was total regeneration and repair of all affected tissues. This observation was made after the histomorphological evaluation of the tissues from the 18-month interim sacrifice.

Formaldehyde-induced lesions in this study were observed in the lining epithelium of the nasal cavity and the proximal trachea. In the nasal cavity,

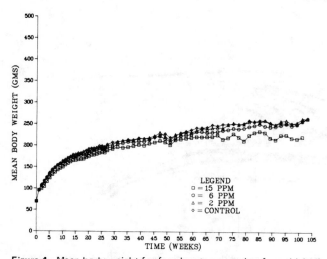

Figure 4 Mean body weight for female rats exposed to formaldehyde vapor.

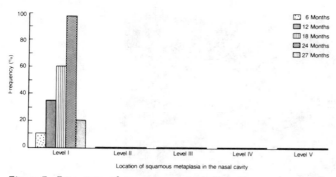

Figure 5 Frequency of squamous metaplasia in the nasal cavity (level I) of Fischer 344 rats exposed to 2 ppm formaldehyde vapor.

lesions were first noted in the anterior sections (levels I, II, and III) from the animals that were terminated at 6 months in the 14.3-ppm exposure group. Alterations of the epithelium in these levels were initially restricted to the ventrolateral portion of the nasal septum and the distal tips of the nasoturbinates and maxilloturbinates. As the study progressed, these lesions increased in severity and distribution in all exposure groups.

In the 2.0-ppm exposure group, squamous metaplasia was present in level I at 12 months (Fig. 5). The mucosa at this location (level I) was characterized by a shift from normal, nonciliated, simple cuboidal epithelium to an epithelial lining that was several cells in thickness and more squamoid in appearance. The organization and the polarity of the individual epithelial cells had changed from vertical to horizontal with respect to the basement membrane. These alterations were termed zones of epithelial dysplasia (Fig. 6). The morphological diagnosis, squamous metaplasia, was used to designate zones of altered epithelium that were

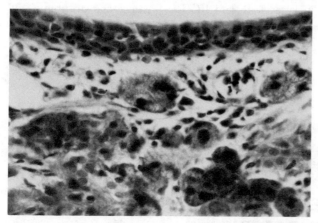

Figure 6 Level I. Epithelial dysplasia in the nasoturbinate of a 15-ppm formaldehyde-exposed rat. H&E, × 400.

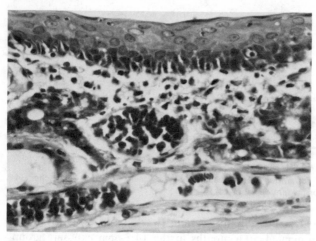

Figure 7 Level I. Squamous metaplasia in the nasoturbinate in a rat exposed to 2.0-ppm formaldehyde vapor for 24 months. H&E, ×400.

characterized by the presence of a well-differentiated germinal cell layer (stratum germinativum) and superficial layers (stratum spinosum and stratum corneum) of epithelium (Fig. 7). Keratin was produced only in areas of metaplastic squamous epithelium. In all exposure groups, epithelial dysplasia was detected earlier than squamous metaplasia. At 24 months in the 2.0-ppm exposure group, the frequency of metaplasia exceeded that of prior sacrifice intervals (Fig. 5). Dysplasia and metaplasia were not observed in levels II, III, IV, or V in this group. At 27 months (3 months post exposure), there was a marked decrease in the frequency of metaplasia in this group (Fig. 5).

In the 5.6-ppm exposure group, epithelial dysplasia and squamous metaplasia were observed in levels I, II, and III. The presence and distribution of metaplasia within the nasal cavity was dependent on the total duration of exposure (Fig. 8). At 27 months, there was regression of squamous metaplasia in

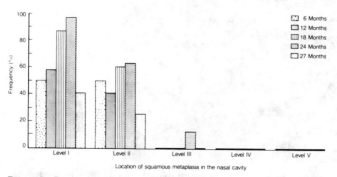

Figure 8 Frequency of squamous metaplasia in the nasal cavity of Fischer 344 rats exposed to 5.6 ppm formaldehyde vapor.

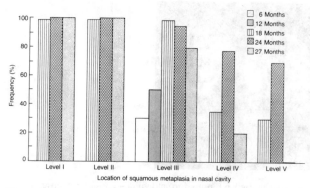

Figure 9 Frequency of squamous metaplasia in the nasal cavity of Fischer 344 rats exposed to 14.3 ppm formaldehyde vapor. At the 6- and 12-month interim sacrifices, levels I, II, IV, and V were not evaluated in this group.

all affected levels (Fig. 8). In the 14.3-ppm exposure group, regression of metaplasia was evident only in levels IV and V (Fig. 9). As in the 2.0-ppm exposure group, the severity of the lesions in the 5.6-ppm and 14.3-ppm exposure groups was most intense in level I; however, in these groups there were also exposure-related compound effects in levels II, III, IV, and V (Figs. 8 and 9). These data are all compared to a lesion (dysplasia or metaplasia) frequency of less than 15 percent for the 0-ppm exposure group, wherein they were only present in level I.

Eight rats (four males and four females) from the low exposure group, six male rats from the intermediate exposure group, and five rats (four males and one female) from the high exposure group had benign proliferative lesions (polypoid adenomas) of the nasal mucosa in levels I, II, or III (Table 5). One control male rat had a similar lesion. In some animals, these lesions were visible grossly

Table 5   A Summary of Neoplastic Lesions in the Nasal Cavity of Fischer 344 Rats Exposed to Formaldehyde Vapor

| | Exposure group and sex | | | | | | | |
| | 0 ppm | | 2 ppm | | 6 ppm | | 15 ppm | |
Diagnosis	M	F	M	F	M	F	M	F
Squamous cell carcinoma	0	0	0	0	1	1	51	52
Carcinosarcoma	0	0	0	0	0	0	1	0
Undifferentiated carcinoma or sarcoma	0	0	0	0	0	0	2*	0
Nasal carcinoma	0	0	0	0	0	0	1*	1
Polypoid adenoma	1	0	4	4	6	0	4	1
Osteochondroma	1	0	0	0	0	0	0	0

*A rat in this group also had a squamous cell carcinoma.

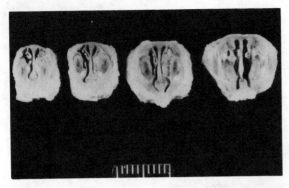

Figure 10 Cross section of a decalcified 5.6-ppm formaldehyde-exposed rat nose with a polypoid adenoma obstructing the right nasal passage. Scale = 1 cm.

in the nasal cavity after decalcification and sectioning (Fig. 10). The tumors grew into the lumen of the nasal cavity (Fig. 11), where, in some cases, they caused obstructive lesions and were associated with focal purulent rhinitis. The cells comprising these neoplasms were cuboidal, rarely ciliated, and often formed acinar-like structures that were filled with detritic cellular and noncellular debris (Fig. 12). The exact origin of the polypoid adenomas (surface or subsurface epithelium) could not be determined by light microscopy. In the control rat, and the rats exposed to 2.0-ppm and 5.6-ppm formaldehyde vapor, the adenomas

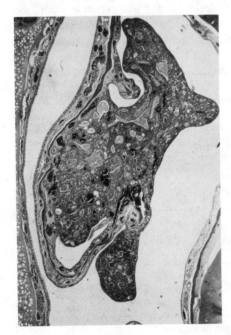

Figure 11 Level II. Macrophotograph of the adenoma viewed in Fig. 10. H&E, X 25.

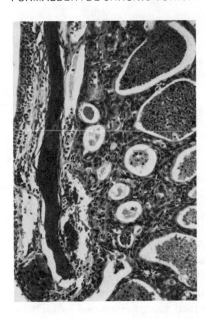

Figure 12 Level II. Polypoid adenoma arising in the nasoturbinate of a rat exposed to 5.6 ppm formaldehyde vapor for 24 months. H&E, × 142.

were usually not associated with zones of epithelial dysplasia or squamous metaplasia.

In the 14.3-ppm exposure group, there were many rats with extensive neoplastic lesions in the nasal cavity. These animals were presented to the necropsy area in a moribund condition characterized by extreme emaciation and severe dyspnea. Many animals were observed at necropsy to have approximately 1 × 1 cm unilateral subcutaneous facial swellings that on closer inspection were interpreted to be proliferative lesions protruding from the nasal cavity (Figs. 13 and 14). On gross examination it was observed that these tumorous lesions

Figure 13 Gross photograph of a 14.3-ppm formaldehyde-exposed rat that has a large invasive squamous cell carcinoma protruding from the nasal cavity. *From Swenberg et al., Cancer Res.* 40:3402 (1980). *Reproduced with permission of Cancer Research.*

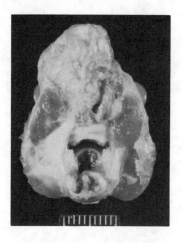

Figure 14 Cross section of an undecalcified 14.3-ppm formaldehyde-exposed rat nose with a destructive squamous cell carcinoma protruding from the nasal cavity. Scale = 1 cm.

originated in the anterior portion of the nasal cavity (levels I, II, or III) and in a few instances they extended into the ethmoturbinates (Fig. 15). Microscopically, the majority were squamous cell carcinomas. As previously described for the 2.0-ppm exposure group in level I, the normal mucosa underwent dysplastic and metaplastic alterations; however, in the 14.3-ppm exposure group, squamous metaplasia was not the endpoint of change. In this group, there was apparently progression of the metaplastic epithelium to zones of squamous epithelial hyperplasia with increased keratin production (Fig. 16) and then to areas of squamous papillary hyperplasia with foci of cellular atypia (Fig. 17). More advanced lesions included carcinoma in situ and invasive squamous cell carcinomas of the nasal turbinates (Fig. 18). The neoplasms were extremely osteolytic and were associated with excessive keratin production and severe purulent rhinitis (Fig. 19). In a few animals, the carcinomas had grown through the ethmoid plate and invaded the meninges of the rhinencephalon. Others grew in an anteroventral direction and invaded the vomeronasal organ but never protruded through the

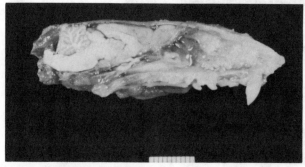

Figure 15 Midsagittal section of a rat nose exposed to 14.3 ppm formaldehyde vapor for 24 months demonstrates carcinoma in the anterior turbinates with extension into the ethmoturbinates. Scale = 1 cm.

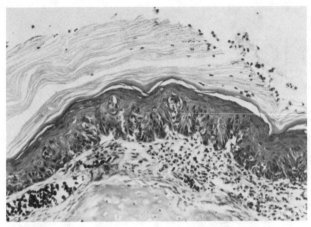

Figure 16 Level II. Squamous epithelial hyperplasia with increased keratin production in the maxilloturbinate of a 5.6-ppm formaldehyde-exposed rat. Note the accumulation of neutrophilic exudate in the lumen. H&E, × 100.

hard palate into the oral cavity. In one rat, there was detectable metastasis of malignant squamous epithelium to the mandibular lymph nodes.

In many animals without carcinoma, excessive accumulation of keratin and inflammatory exudate caused severe dyspnea and death. This is not unexpected as nasotracheal ventilation in rats is obligatory.

Squamous cell carcinomas were observed in two rats exposed to 5.6 ppm and in 103 rats from the 14.3-ppm exposure group. A carcinosarcoma, one undifferentiated carcinoma, and an undifferentiated sarcoma were also observed in the 14.3-ppm exposure group (Table 5). In addition to the squamous cell

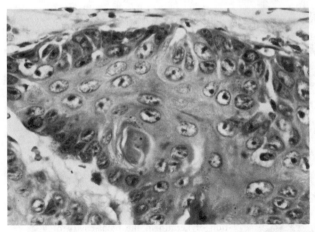

Figure 17 Level II. Nest of cellular atypia in a papillary frond of metaplastic epithelium from a 14.3-ppm formaldehyde-exposed rat. H&E, × 400.

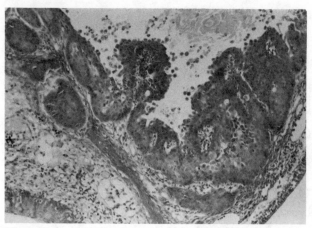

Figure 18 Level I. Early invasive squamous cell carcinoma in the maxilloturbinate of a rat exposed to 5.6 ppm formaldehyde vapor for 24 months. H&E, ×63.

carcinomas, there were two animals with carcinomas of the respiratory epithelium (Table 5).

Of the 106 rats with nasal neoplasia, 83 were recorded as unscheduled deaths. The sharp decrease in cumulative survival in the 14.3-ppm exposure group began after 12 months of exposure and continued to 24 months (Figs. 20 and 21). This was primarily associated with the development of nasal carcinomas (Fig. 22).

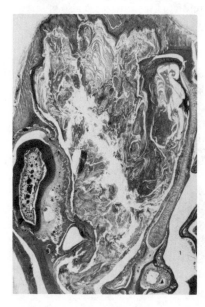

Figure 19 Level II. Advanced squamous cell carcinoma with excessive accumulation of keratin in the nasal cavity of a 14.3-ppm formaldehyde-exposed rat. H&E, ×10.

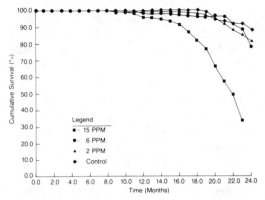

Figure 20   Cumulative survival of male rats exposed to formaldehyde vapor.

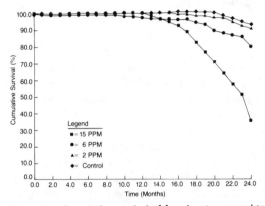

Figure 21   Cumulative survival of female rats exposed to formaldehyde vapor.

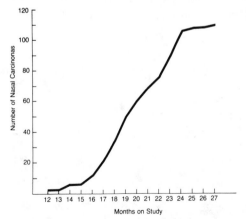

Figure 22   Cumulative number of nasal carcinomas in rats exposed to 14.3 ppm formaldehyde vapor. Exposure was terminated at 24 months.

In the rats exposed to 14.3 ppm that were killed at 18 months, there were a few animals that had multifocal areas of minimal to mild epithelial hyperplasia, epithelial dysplasia, or squamous metaplasia of the proximal tracheal mucosa. These alterations in the epithelium were morphologically similar to those visualized in the anterior sections (levels I and II) of the nasal cavity from the 2.0-, 5.6-, and 14.3-ppm exposure groups. Similar lesions were also observed with an increased frequency in the group of unscheduled deaths and at the 24-month sacrifice. Tracheal lesions were not observed at 27 months in the 14.3-ppm exposure group. There were no significant tracheal lesions present in the 0-ppm, 2.0-ppm, or 5.6-ppm exposure groups. Neoplasia of the tracheal epithelium was not observed.

Exposure to formaldehyde also produced a dose-related increase in yellow discoloration of the haircoat. Three months after the exposures were discontinued, haircoat coloration was essentially normal.

Mice

From 6 months on, there was an apparent exposure-related decrease in cumulative survival among male mice (Fig. 23), while the female mice demonstrated little change from controls until the fourth quarter of the inhalation exposure period (Fig. 24). There were no apparent differences in body weight throughout the study for either sex (Figs. 25 and 26).

Significant formaldehyde-induced lesions in mice were also observed in the upper respiratory tract. As in the rat, neoplasia appeared to be preceded by dysplastic and metaplastic alterations of the nasal mucosa in the 14.3-ppm exposure group. Lesions in the nasal cavity of the mice were first detected at 12 months, when animals in the 14.3-ppm exposure group exhibited serous rhinitis in all evaluated levels. By the 18-month sacrifice, many animals in the 14.3-ppm exposure group had dysplastic and metaplastic alterations of the nasal

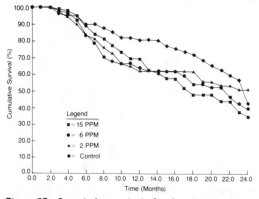

Figure 23 Cumulative survival of male mice exposed to formaldehyde vapor.

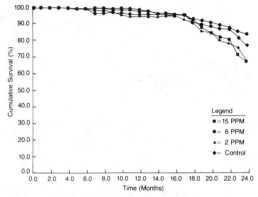

Figure 24   Cumulative survival of female mice exposed to formaldehyde vapor.

mucosa in levels II and III (Fig. 27). These changes were associated with a shift in the nasal exudate from serous to purulent. In the 5.6-ppm exposure group at the 18-month sacrifice, a few mice had dysplastic changes that were associated with serous rhinitis in level II, and no alterations were detected in the nasal cavity from animals in the 2.0-ppm exposure group.

By 24 months, a majority of the mice in the 14.3-ppm exposure group had dysplastic and metaplastic alterations that were associated with seropurulent rhinitis in all animals. At this time period, there were also only a few mice in the 5.6-ppm exposure group with dysplasia, metaplasia, or serous rhinitis. At 24 months, the mice exposed to 2.0 ppm were still free of degenerative lesions of the mucosa; however, a few animals had serous rhinitis in level II.

Two male mice (14.3 ppm) from the 24-month sacrifice had squamous cell

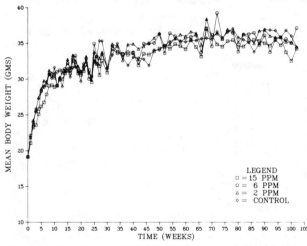

Figure 25   Mean body weight for male mice exposed to formaldehyde vapor.

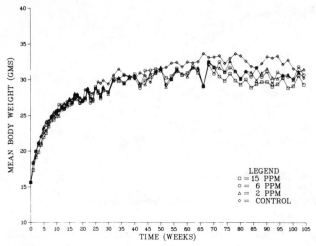

Figure 26 Mean body weight for female mice exposed to formaldehyde vapor.

carcinomas in the nasal cavity (Fig. 28). Both carcinomas originated unilaterally on the nasoturbinates. The location and the histomorphology of these tumors were similar to those observed in rats; however, they were not as invasive and did not cause death.

At 24 months there were also mice in all exposure groups with minimal to moderate hyperplasia of the squamous epithelium lining the nasolacrimal duct. This lesion was most severe, both in frequency and distribution, in mice from the 14.3-ppm exposure group. Animals from this group also had focal areas of atrophy of the epithelium lining the ethmoturbinates. This lesion also occurred in the 5.6-ppm group, but the frequency was greatly reduced. Tracheal lesions were not observed in mice.

At 27 months, dysplastic epithelial lesions were present only in the 14.3-

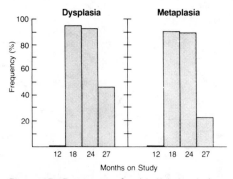

Figure 27 Frequency of epithelial dysplasia and squamous metaplasia in the nasal cavity (level II) of B6C3F1 mice exposed to 14.3 ppm formaldehyde vapor.

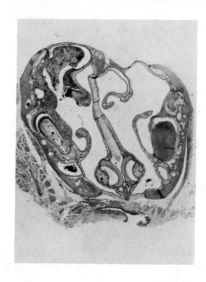

Figure 28 Level II. Macrophotograph of an advanced squamous cell carcinoma in a B6C3F1 mouse exposed to 14.3 ppm formaldehyde vapor for 24 months. H&E, ×10.

ppm exposure group (Fig. 27), and the exudate associated with these lesions was more serous than purulent. Metaplasia was not present at this time interval. The remaining formaldehyde-exposed groups were free of significant pathology at this time interval.

DISCUSSION

The formaldehyde-induced nasal lesions in mice were much less severe when they were compared with similar lesions in rats from the same exposure group. This apparent discrepancy in exposure-related response between the two species may be related to differences in their physiological responses to irritant inhalation (JA Swenberg, personal communication). Therefore, it is noteworthy to mention that the frequency of carcinoma formation in the mice exposed to 14.3 ppm was similar to the frequency of carcinoma formation in the rats exposed to 5.6 ppm.

In the mice, as in rats, there was a progressive increase in the severity, distribution, and frequency of the irritant-induced lesions (dysplasia or metaplasia) in the nasal cavity when the presence of these lesions is compared to their temporal development and exposure concentration (Figs. 5, 8, 9, and 27).

CONCLUSION

The inhalation of formaldehyde vapor for 24 months was associated with an exposure-related increase in the frequency, severity, and distribution of epithelial dysplasia and squamous metaplasia of the respiratory epithelium lining the anterior portion of the nasal cavity in rats from all exposure groups. In contrast to rats, mice exhibited only dramatic irritant-induced effects (epithelial dysplasia, squamous metaplasia, or both) at the highest exposure level and essentially no effect at the lowest exposure level.

Three months after formaldehyde exposure was discontinued in rats, there was apparent regression of metaplasia in all affected levels of the nasal cavity in both the low and intermediate exposure groups and levels IV and V in the high exposure group. In mice, regression of metaplasia was evident in all affected levels of the intermediate and high exposure groups.

Squamous cell carcinomas were observed in the nasal cavity of male mice (14.3 ppm) and rats (5.6 ppm and 14.3 ppm) of both sexes following the inhalation of formaldehyde for 2 years. Formaldehyde inhalation in rats was also associated with an apparent increase in the frequency of a previously unreported spontaneous neoplasm in the nasal cavity—polypoid adenoma. This benign tumor was present in all groups.

A complete evaluation of formaldehyde carcinogenicity and the determination of a no-effect level in rats and mice must await adequate statistical evaluation and the completion of this and other related studies.

REFERENCES

1 *Directory of Chemical Producers: United States of America*, pp. 637–638. Menlo Park: SRI International (1979).

2 U.S. Department of Health, Education, and Welfare, Public Health Service, Center for Disease Control, National Institute for Occupational Safety and Health. *Criteria for a Recommended Standard. Occupational Exposure to Formaldehyde.* DHEW (NIOSH) Publication No. 77-126. Washington, D.C.: U.S. Government Printing Office (1976).

3 *Formaldehyde and Other Aldehydes.* Washington, D.C.: National Academy Press (1981).

4 Andersen, I. Formaldehyde in the indoor environment—health implications and the setting of standards, pp. 65–77, and discussion, pp. 77–87. In PO Fanger and O Valbjorn (eds.), *Indoor Climate. Effects on Human Comfort, Performance, and Health in Residential, Commercial, and Light-Industry Buildings.* Proceedings of the First International Indoor Climate Symposium, Copenhagen, August 30–September 1, 1978. Copenhagen: Danish Building Research Institute (1979).

5 Odom, RB, and HI Maibach. Contact urticaria: A different contact dermatitis. In FN Marzulli and HI Maibach (eds.), *Advances in Modern Toxicology. Vol. 4. Dermatotoxicology and Pharmacology*, pp. 441–453. Washington, D.C.: Hemisphere Publishing Corporation (1977).

6 North American Contact Dermatitis Group. Epidemiology of contact dermatitis in North America: 1972. *Arch. Dermatol.* 108:537–540 (1973).

7 Sellakumar, AR, RE Albert, GM Rusch, GV Katz, N Nelson, and M Kuschner. Inhalation carcinogenicity of formaldehyde and hydrogen chloride in rats. *Proc. Am. Assoc. Cancer Res.* 21:106 (1980).

8   Auerbach, C, M Moutschen-Dahmen, and M Moutschen. Genetic and cyto-
genetical effects of formaldehyde and related compounds. *Mutat. Res.*
39:317–362 (1977).

9   Martin, CN, AC McDermid, and RC Garner. Testing of known carcinogens
and noncarcinogens for their ability to induce unscheduled DNA synthesis
in HeLa cells. *Cancer Res.* 38:2621–2627 (1978).

10   Wilkins, RJ, and HD MacLeod. Formaldehyde induced DNA-protein cross-
links in *Escherichia coli. Mutat. Res.* 36:11–16 (1976).

11   Hsie, AW, JP O'Neill, JR San Sebastian, DB Couch, PA Briner, WNC Sun,
JC Fuscoe, NL Forbes, R Machanoff, JC Riddle, and MH Hsie. Quantitative
mammalian cell genetic toxicology: Study of the cytotoxicity and mutagenicity of seventy individual environmental agents related to energy technologies and three subfractions of a crude synthetic oil in the CHO/-HGPRT
system. In MD Waters, S Nesnow, JL Huisingh, SS Sandhu, and L Claxton
(eds.), *Application of Short-Term Bioassays in the Fractionation and Analysis of Complex Environmental Mixtures*, pp. 293–315. New York: Plenum
Press (1978).

12   Obe, G, and B Beek. Mutagenic activity of aldehydes. *Drug Alcohol Depend.*
4:91–94 (1979).

Chapter 12

Mechanisms of Formaldehyde Toxicity

James A. Swenberg

Elizabeth A. Gross

Joseph Martin

James A. Popp

The discovery that exposure to formaldehyde vapor could induce nasal carcinomas in rodents with striking species differences and an extremely sharp dose
response (1, 2) stimulated the development of a series of research projects aimed
at understanding the mechanisms involved in formaldehyde toxicity (see also
Chap. 11, this volume). This presentation will review various aspects of tissue
exposure and response, using preliminary data from investigations designed to
fill some of the gaps in our knowledge of the pathogenesis of formaldehyde-
induced disease.

STUDIES ON THE DISTRIBUTION OF FORMALDEHYDE
IN THE NASAL CAVITY

Biochemical investigations on the absorption and distribution of [^{14}C] formaldehyde (CH$_2$O) have demonstrated that it is primarily deposited in the upper

The authors would like to thank Drs. Craig Barrow and Kevin Morgan for their interest
and support of this research and for thoughtful criticism of the manuscript. They are also
grateful to Ms. Holly Randall for her expertise in preparing the tissues for histological examination, Ms. Mary Bedell for her assistance with the autoradiographic studies, and Ms.
Belinda Scortichini for assistance with the Videoplan.

respiratory tract (see Chap. 4, this volume). In order to localize this deposition within the nasal cavity, naive or pretreated rats and mice were exposed to 15 ppm [^{14}C] formaldehyde for 6 h. The animals were killed by decapitation immediately after removal from the chambers. The skin was removed from the heads, which were then frozen in carboxymethylcellulose for cryosectioning in an LKB PMV 2250. Autoradiographic films were prepared from transverse step sections of the nasal cavity using NS2T emulsion X-ray films (Eastman Kodak Co.) as previously described (3).

Formaldehyde-associated radioactivity was heavily deposited in the anterior nasal cavity of rats and mice. The localization of radioactivity correlated well with the distribution of lesions in similarly exposed animals, that is, activity was greatest in regions of respiratory epithelium over the maxilloturbinates and nasoturbinates and the lateral wall of the nasal cavity (Fig. 1a). An exception was noted for the ventral portion of the nasal cavity, which is lined with squamous epithelium. Radioactivity was heavily deposited in this area; however, minimal histologic evidence of toxicity was observed. This suggests that squamous epithelium is less sensitive to formaldehyde intoxication than is respiratory epithelium. The induction of squamous metaplasia as a response to CH_2O exposure and the relative resistance of squamous epithelium to toxicity suggest

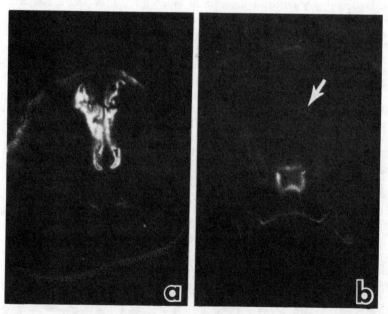

Figure 1 Autoradiograph of transverse sections of the nasal cavity from a rat exposed to 15 ppm [^{14}C] formaldehyde for 6 h. Note the heavy deposition of radioactivity in the anterior portion (a) and the paucity of radioactivity in the region of olfactory epithelium (X) compared to moderate amounts of radioactivity in the nasopharynx of a posterior section (b) of nasal cavity.

that squamous metaplasia is a host defense mechanism to formaldehyde toxicity. A portion of the radioactivity located in the ventral squamous regions may be the result of mucociliary flow and gravity rather than direct exposure to formaldehyde vapor. Radioactivity may represent material covalently bound to mucus rather than reactive formaldehyde. The amount of radioactivity deposited in regions of olfactory mucosa in the nasal cavity was much less; thus radioactivity in posterior sections was primarily confined to the nasopharynx (Fig. 1b). Whole-body parasagittal sections of rats confirmed the anterior-posterior concentration gradient of [^{14}C] formaldehyde but demonstrated the presence of radioactivity down to the level of the bronchial lining (Fig. 2). Bronchial radioactivity was similar to that of bone marrow and is thought to be the result of metabolic incorporations of the one-carbon pool. Minimal differences were apparent in formaldehyde distribution between naive rats and mice. When animals that had been exposed to 15 ppm of nonradioactive formaldehyde for 9 days prior to exposure to [^{14}C] formaldehyde were compared to naive rats, a decrease in radioactivity in the dorsal olfactory epithelium of the anterior half of the nasal cavity and of the abdominal viscera was noted. Changes in airflow due to vascular congestion and inflammation may be responsible for decreases in nasal cavity deposition, while the decreased radioactivity associated with the abdominal viscera may reflect a decrease in mucociliary flow and subsequent swallowing of radioactive mucus. Sharp anterior-posterior concentration gradients, such as demonstrated in these investigations, might be expected for reactive chemicals with a high degree of water solubility.

MORPHOLOGICAL STUDIES ON THE NASAL CAVITY

Since differences in airway volume, shape, and surface area could readily affect response patterns, we have undertaken a series of investigations on the normal

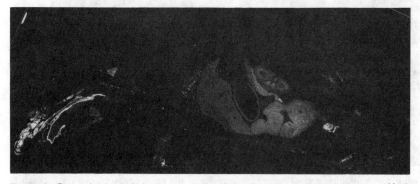

Figure 2 Parasagittal whole-body autoradiograph of a rat exposed to 15 ppm [^{14}C] formaldehyde for 6 h. Note the heavy deposition of radioactivity in the nasal cavity and anterior half of the trachea, with additional moderate deposition in the bronchi of the lungs. Radioactivity in the abdominal viscera is probably the result of mucociliary clearance and metabolites.

structure of the rat and mouse nasal cavity using morphometric analysis. Conventional surface areas, volumes, and the distribution of the various epithelial types lining the nasal cavity in normal 7- and 16-week-old Fischer 344 male rats and B6C3F1 male mice have been mapped at the light microscopic level (4). Photographs of nasal transverse sections were analyzed using a Zeiss Videoplan computerized image analysis system programmed for measurement and evaluation of count, area, perimeter, and length. The results are shown in Table 1. The percentage of the surface area covered by the various epithelial types lining the nasal cavity was similar in all animals studied. Little change in volume or surface area was present between 7- and 16-week-old mice, whereas the volume and surface area of the rat nasal cavity increased 165 percent and 168 percent, respectively. Utilizing the autoradiographic data on patterns of formaldehyde deposition, the morphometry provides baseline data that can be used to help quantitate the "dose" of inhaled chemical reaching the nasal cavity and may be useful in understanding differences in species' response to the same concentration of inhaled chemical. For example, using minute volumes for rats and mice exposed to 15 ppm formaldehyde vapor (5; see also Chap. 3, this volume), the dose of formaldehyde available for absorption is 0.154 and 0.075 $\mu g/min/cm^2$ in rats and mice, respectively. Thus, the mouse nasal mucosa is exposed to only half the amount of formaldehyde that the rat nasal mucosa is exposed to. This dose correlates well with tumor data, in which the incidence of nasal carcinoma is similar in rats exposed to 6 ppm and mice exposed to 15 ppm of formaldehyde vapor (see Chap. 11, this volume). Morphometric analysis of the nasal cavity could also be used to quantitate changes in epithelial types (e.g., squamous metaplasia) that occur in subchronic and chronic inhalation studies. One must bear in mind, however, that surface areas determined by light microscopy are dramatically different from actual cellular surface areas when the epithelium is composed of cells with cilia and microvilli. Measurements derived by light microscopic morphometry do, however, closely approximate the surface area of the mucous blanket that normally lines the nasal cavity and provides the initial site for formaldehyde absorption.

Preliminary transmission and scanning electron microscopic studies of the rat nasoturbinate have identified nine different cell types within the respiratory epithelium. Nasal epithelial cells become progressively taller from anterior to posterior. The anterior cells (behind the normal squamous epithelium) have few surface microvilli, while cells farther posterior have more numerous microvilli (Fig. 3). Columnar nonciliated epithelial cells having a large amount of smooth endoplasmic reticulum in the apical cytoplasm may represent more metabolically active cell types (Fig. 4). The percentage of cells having apical cilia increases from anterior to posterior (Fig. 5). Such cells are thought to play a major role in normal mucociliary clearance mechanisms and may represent a potential target for formaldehyde intoxication. Changes in blood flow and congestion are likely to cause alterations in airflow, which in turn could

Table 1 Morphometric Analysis of the Rat and Mouse Nasal Cavity

	Rat		Mouse	
	7 Wk (115 g)	16 Wk (288 g)	7 Wk (30 g)	16 Wk (33 g)
Length (mm)	$7.3 \pm 0.0^*$	9.1 ± 0.3	4.9 ± 0.7	5.1 ± 0.08
Volume (mm³)	155.5 ± 1.3	256.7 ± 4.1	32.5 ± 3.2	31.5 ± 2.1
Surface area (mm²)				
Squamous epithelium (mm²)	27.7 ± 1.9	44.2 ± 5.2	20.9 ± 0.4	20.6 ± 2.2
Respiratory epithelium	352.4 ± 4.9	623.1 ± 14.0	132.4 ± 5.7	133.9 ± 4.6
Olfactory epithelium	418.5 ± 19.2	675.2 ± 43.0	125.5 ± 4.0	136.9 ± 7.3
Total surface area	798.6 ± 20.2	1343.5 ± 55.0	277.7 ± 16.1	289.0 ± 13.1

Note: From Gross et al. (9).
*Values are expressed as mean of 3 animals ± standard deviation.

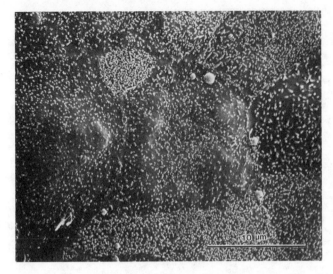

Figure 3 Scanning electron micrograph of anterior respiratory epithelial cells with microvilli.

alter the extent and site of toxicity. The extensive number of blood vessels and sinuses can be readily appreciated by examining perfused specimens with light or electron microscopy (Fig. 6).

ACUTE TOXICITY FOLLOWING FORMALDEHYDE EXPOSURE

Acute cell degeneration, necrosis, and inflammation were evident in the nasal cavities of rats exposed to 15 ppm formaldehyde vapor for 1–9 days (6 h/day).

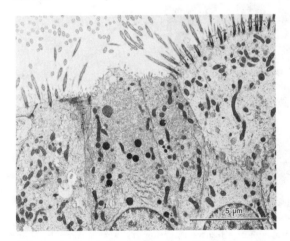

Figure 4 Transmission electron micrograph of a columnar cell with considerable smooth endoplasmic reticulum.

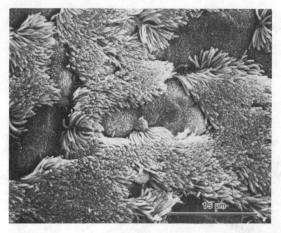

Figure 5 Scanning electron micrograph of ciliated respiratory epithelial cells from the nasoturbinate.

Initial lesions were most severe on the tips of the maxilloturbinates and naso-turbinates. Acute degeneration of the respiratory epithelium, with edema and congestion, was evident at the end of 1 day's exposure (Fig. 7). This was fol-lowed by ulceration, necrosis, and an influx of inflammatory exudate at days 3-9 (Fig. 8). Early squamous metaplasia was detected over the nasoturbinates and maxilloturbinates, median septum, and lateral wall after as little as 5 days of formaldehyde exposure (Fig. 9). Examination of turbinates from rats exposed 5 days and allowed to recover for 48 h demonstrated considerable regeneration. Areas that were frequently ulcerated, such as the lateral wall, had single thin strap cells covering areas normally occupied by three or more cuboidal epithelial

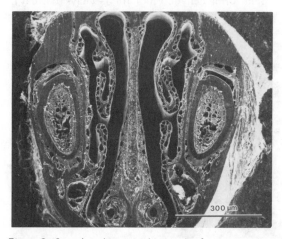

Figure 6 Scanning electron micrograph of a transverse section of the rat nasal cavity. Note the large number of blood vessels just under the epithelial surfaces.

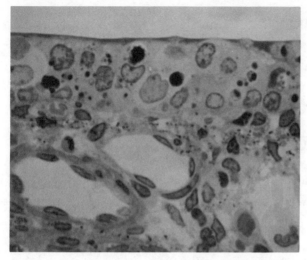

Figure 7 A 1 μm plastic section of a perfused nasoturbinate from a rat exposed to 15 ppm formaldehyde for 6 h. Note the early degeneration and swelling. X 1375.

cells (Fig. 10). In contrast to these changes in the respiratory epithelium, mild serous rhinitis was the principal lesion in regions of olfactory epithelium. Mild degenerative and inflammatory changes were also evident in the nasopharynx.

The mouse was similar, but less severe than the rat, in its acute response to formaldehyde exposure. Five days' exposure to 15 ppm formaldehyde vapor caused degeneration, focal necrosis, and inflammation of the nasoturbinates

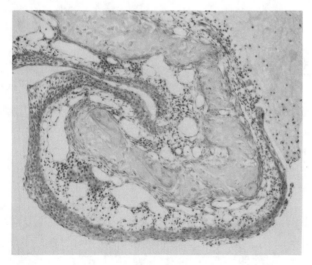

Figure 8 Focal ulceration, epithelial hyperplasia, necrosis, and rhinitis of the nasoturbinate of a rat exposed to 15 ppm formaldehyde for 3 days (6 h/day). X 152.

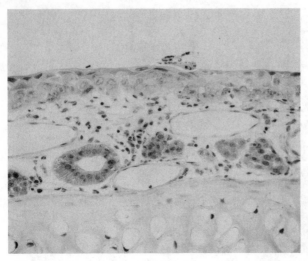

Figure 9 Early squamous metaplasia of the respiratory epithelium after 5 days' exposure to 15 ppm formaldehyde. X 304.

and maxilloturbinates and to the lateral walls, but there was minimal toxicity to areas lined by squamous or olfactory epithelium.

By comparing these acute toxicity studies with data from the 3-month exposures reported by Rusch (see Chap. 10, this volume) and the interim sacrifice data from the formaldehyde bioassay (see Chap. 11, this volume), it is clearly evident that adaptive changes occur. The extent and severity of formaldehyde-

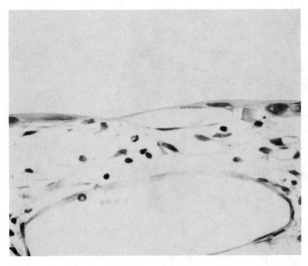

Figure 10 A region of the lateral wall is covered by single "strap" cells following 5 days' exposure to 15 ppm (6 h/day) and 2 days recovery. X 608.

induced toxicity diminished with time. This may be due to changes in respiratory physiology, as well as to alterations at the cellular level, such as squamous metaplasia, epithelial hyperplasia, and increased detoxification pathways.

EFFECTS OF FORMALDEHYDE EXPOSURE
ON CELL TURNOVER

A prominent response to cell loss associated with toxicity is compensatory cell replication. Surviving cells undergo division in order to replace dead cells and to increase the thickness of the epithelium. Preliminary studies on the effect of 15 ppm of formaldehyde vapor, 6 h/day for 5 days, have been conducted in rats. One day prior to exposure the animals had an osmotic mini-pump containing a 7-day supply of ^{3}H-thymidine implanted in the peritoneal cavity. Following the fifth day of exposure the animals were anesthetized with pentobarbital and killed by vascular perfusion with 10% neutral buffered Formalin; the heads were decalcified, embedded, and sectioned and slides were prepared for autoradiography. Control rats had 5/2686 (0.19 percent) labeled respiratory epithelial cells, while formaldehyde-exposed animals had 634/4712 (13.5 percent), which represents a 71-fold increase in cell replication. Sequential pulse labeling studies following 1, 3, 5, or 9 days of formaldehyde exposure demonstrated that maximum cell proliferation occurred after 3 days. A 10- to 20-fold increase in cell replication occurred when rats were exposed for 3 days to 6 or 15 ppm formaldehyde vapor (Fig. 11) and when mice were exposed to 15 ppm. Similar

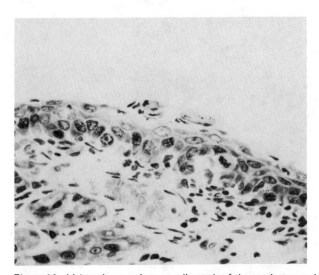

Figure 11 Light microscopic autoradiograph of the respiratory epithelium of a rat exposed to 6 ppm formaldehyde for 3 days. Note the large number of cells labeled with ^{3}H-thymidine. ✕ 608.

exposures of rats to 0.5 or 2.0 ppm and mice to 0.5, 2.0, or 6.0 ppm formaldehyde (Fig. 12) did not result in increased cell turnover in the nasal cavity.

MACROMOLECULAR INTERACTIONS
WITH FORMALDEHYDE

The initiating event in chemical carcinogenesis is thought to involve covalent
binding of reactive compounds to specific sites on the DNA (6). Such damage
can be enzymatically removed by cellular DNA repair systems. Furthermore,
cell division is required prior to DNA repair for carcinogenesis to occur. In the
case of formaldehyde, unwinding of the double helix during cell replication
may also be required to expose critical sites on the DNA to covalent binding
(7). CH_2O does not react with native double-stranded DNA, but it does react
with denatured DNA (8). Denaturation destroys H-bonding between two strands
of the molecule similar to the unwinding of DNA during replication. This
phenomenon could explain the apparent cell-cycle specificity of CH_2O. Mutagenesis in *Drosophila* is restricted to the period of chromosome replication pre-
ceding meiosis (8). Exponentially growing cultures of yeast exhibit greater
sensitivities to lethality and mutagenesis than do stationary cultures (9). Carcinogenesis in the rat nasal epithelium may be directly related to the increased
cell division resulting from toxicity. During the increased cell division, the likelihood of interaction of CH_2O with DNA would increase, as would fixation of
adducts before DNA repair could occur.

   The reaction of CH_2O with macromolecules occurs largely by way of amino

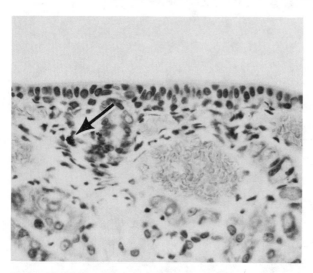

Figure 12 Light microscopic autoradiograph from a rat exposed to 2 ppm formaldehyde
for 3 days. Note the lack of ^{3}H-thymidine labeled epithelial cells. A labeled stromal cell
is present in the submucosa (*arrow*). X 608.

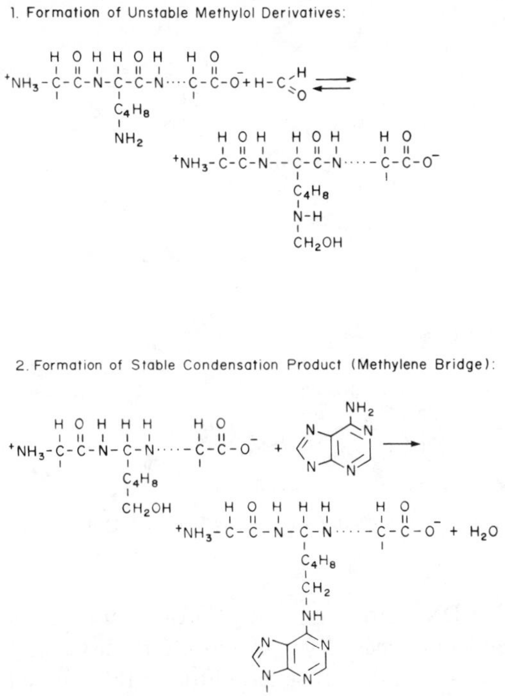

1. Formation of Unstable Methylol Derivatives:

2. Formation of Stable Condensation Product (Methylene Bridge):

Figure 13 Chemical reaction thought to be involved in protein-DNA cross-links due to formaldehyde.

groups, such as those in proteins and nucleic acids. A two-step mechanism has been proposed (10), the first step of which is the fast, reversible formation of unstable methylol derivatives which may lose water to form a methylene Schiff base. The irreversible formation of a stable condensation product, a methylene cross-link, may occur by way of nucleophilic attack on the methylene carbon. These cross-links could occur between two proteins, between DNA and protein (Fig. 13), or between DNA and DNA (Fig. 14). By involvement of DNA, the latter two processes could constitute the mechanism by which an irreversible genetic change occurs. Biological implications of such reactions have been demonstrated by comparing DNA repair deficient and proficient strains of *Escherichia coli*, yeast, and mammalian cells (8, 11).

One of the most useful techniques for evaluating DNA-protein and DNA-DNA cross-links is alkaline elution (12, 13). This technique measures the amount of DNA retained on a filter versus the amount of DNA eluting through the filter (Fig. 15). Typically, about 90 percent of control DNA from tissues or cultured cells is retained on the filter. When single-strand breaks are present, the amount of DNA eluting increases proportional to the amount of damage.

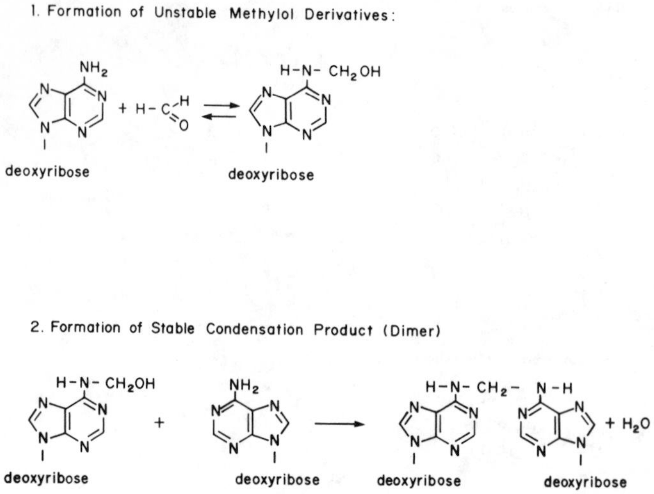

1. Formation of Unstable Methylol Derivatives:

NH₂
+ H - C ⟷ H - N - CH₂OH
deoxyribose   deoxyribose

2. Formation of Stable Condensation Product (Dimer)

H - N - CH₂OH   NH₂   →   H - N - CH₂ - N - H   + H₂O
deoxyribose   deoxyribose   deoxyribose   deoxyribose

Figure 14 Possible reactions involved in the formation of DNA-DNA dimers by formaldehyde.

If, however, protein-DNA or DNA-DNA cross-links exist, DNA elution is decreased. By treating the DNA with proteinase K, it is possible to distinguish between protein-DNA and DNA-DNA cross-links. We have utilized this method to examine the interactions of formaldehyde and DNA.

In order to increase the sensitivity of the assay for cross-linked DNA, single-strand breaks or alkali-labile sites were induced in control and treated samples. V79 cells were exposed to 0, 60, 120, or 600 μmols of formaldehyde solution for 1 h followed by 15 μmols of N-methyl-N'-nitro-N-nitroso-guanidine (MNNG) for 15 min to induce a uniform number of single-strand breaks and alkali-labile sites in the DNA. DNA was eluted from the filters after lysing the cells in the presence or absence of proteinase K. A dose-dependent increase in retention of DNA on the filters was associated with formaldehyde exposure (Fig. 16). This was reversed by proteinase K, providing strong evidence for protein-DNA cross-links. Cells treated with 120 μmols formaldehyde and allowed to grow for 24 h before exposure to MNNG did not show increased DNA retention on the filter, indicating that DNA repair processes had removed the cross-links. Recent studies

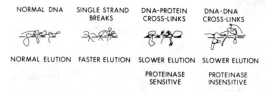

NORMAL DNA	SINGLE STRAND BREAKS	DNA-PROTEIN CROSS-LINKS	DNA-DNA CROSS-LINKS
NORMAL ELUTION	FASTER ELUTION	SLOWER ELUTION	SLOWER ELUTION
		PROTEINASE SENSITIVE	PROTEINASE INSENSITIVE

Figure 15 Schematic on the effect of various forms of DNA damage on the alkaline elution assay.

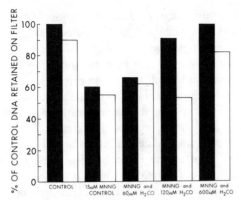

Figure 16 Effect of formaldehyde on the alkaline elution of MNNG-treated V79 cells in the presence (open bars) and absence (closed bars) of proteinase K.

by Ross and Shipley confirmed these findings in L1210 cells and demonstrated that DNA-protein cross-links occur at nonlethal concentrations of formaldehyde (14, 15). Similar results have been reported for yeast (16).

Reaction between CH_2O and DNA has been shown to occur with bases, nucleotides, nucleosides, synthetic polynucleotides, and intact DNA. Using high performance liquid chromatography, cross-linked nucleosides were isolated and identified from formaldehyde-treated DNA (17). Methylene-bridged products included dCyd-CH_2-dGuo, dGuo-CH_2-dGuo, dGuo-CH_2-dCyd, dGuo-CH_2-dAdo, and dAdo-CH_2-dAdo. The relative amounts and biological significance of these DNA-DNA cross-links remain to be determined, however, since they were only demonstrated after reacting formaldehyde and calf thymus DNA for 40 days. Although the products of chemical reactions between CH_2O and DNA bases have been isolated and characterized, no such reaction product has been isolated from formaldehyde-treated cells or tissues.

DISCUSSION

It should be obvious from the data presented above, and elsewhere in this volume, that mechanisms responsible for formaldehyde carcinogenesis are complex. Formaldehyde-induced squamous cell carcinomas of the nasal cavity exhibit an extremely sharp dose-response curve and prominent species differences. Understanding the pathogenesis of these differences is crucial for proper risk assessment. While a great deal of additional research is necessary, it is possible to begin piecing together some of the prominent mechanisms on which the outcome depends. First, it is clear that areas of contact, deposition, and toxicity are similar in rats and mice. These endpoints are dependent on the species' respiratory physiology and response to acute versus chronic exposure. Using such data it is possible to calculate the dose received by the target organ.

By incorporating this information into the bioassay data, the major species difference in dose response disappears, since mice and rats receiving a similar dose developed a similar low incidence of squamous cell carcinomas. Furthermore, similar increases in cell replication were induced by exposure of rats to 6 ppm and mice to 15 ppm formaldehyde. Since cell replication can be a compensatory response to toxicity and may have profound effects on the availability of covalent binding sites on the DNA, toxic exposure levels may greatly enhance covalent binding. Since protein-DNA cross-links represent the major DNA adduct and they can be rapidly repaired, increased cell proliferation may also be important in fixing the mutational events thought to be responsible for initiation. Nothing is presently known about formaldehyde as a promoter; however, increases in cell turnover may accelerate the process of carcinogenesis. The fact that only exposure concentrations associated with squamous cell carcinoma in rats and mice resulted in increased cell proliferation lends strong support to the hypothesis that increased cell proliferation is a critical event in formaldehyde carcinogenesis.

A separate but related consideration is the relative importance of concentration versus cumulative dose. Cytotoxicity and increased cell proliferation appear to be related to the former. If one compares the cumulative dose to rats exposed to 15 ppm formaldehyde 6 h/day, 5 days/wk (450 ppm-h/wk) with that of rats exposed to 3 ppm for 22 h/day, 7 days/wk (462 ppm-h/wk) it is readily apparent that similar cumulative doses of formaldehyde resulted, yet the lesions (see Chaps. 10 and 11, this volume) and, presumably, the amount of cell proliferation are vastly different. It would appear then that formaldehyde concentration is the most important consideration in determining response and that the response changes drastically when toxic concentrations are achieved. Additional research is needed to better define the concentration-response relationships associated with formaldehyde exposure, as these data will be crucial for cogent risk assessment.

REFERENCES

1 Swenberg, JA, WD Kerns, RI Mitchell, EJ Gralla, and KL Pavkov. Induction of squamous cell carcinomas of the rat nasal cavity by inhalation exposure to formaldehyde vapor. *Cancer Res.* 40:3398–3402 (1980).

2 Swenberg, JA, WD Kerns, KL Pavkov, RI Mitchell, and EJ Gralla. Carcinogenicity of formaldehyde vapor: Interim findings in a long-term bioassay of rats and mice. In *Mechanisms of Toxicity and Hazard Evaluation*, edited by B Holmstedt, R Lauwerys, M Mercier, and M Roberfroid, pp. 283–286. Amsterdam: Elsevier (1980).

3 Irons, RD, and EA Gross. Standardization and calibration of whole-body autoradiography for routine semiquantitative analysis of the distribution of ^{14}C-labelled compounds in animal tissues. *Toxicol. Appl. Pharmacol.* 59: 250–256 (1981).

4 Gross, EA, JA Swenberg, S Fields, and JA Popp. Comparative morphometry of the nasal cavity of rats and mice. *J. Anat.* (in press).

5 Chang, JCF, WH Steinhagen, and CS Barrow. Effect of single or repeated formaldehyde exposure on minute volume of B6C3F1 mice and F-344 rats. *Toxicol. Appl. Pharmacol.* 61:451–459 (1981).

6 Pegg, AE. Formation and metabolism of alkylated nucleosides: Possible role in carcinogenesis by nitroso compounds and alkylating agents. *Adv. Cancer Res.* 25:195–269 (1977).

7 Singer, B, and JT Kusmierek. Chemical mutagenesis. *Ann. Rev. Biochem.* 52: 655–692 (1982).

8 Auerbach, C, M Moutschen-Dahmen, and J Moutschen. Genetic and cyto-genetical effects of formaldehyde and related compounds. *Mutat. Res.* 39: 317–362 (1977).

9 Chanet, R, and RC von Borstel. Genetic effects of formaldehyde in yeast. III. Nuclear and cytoplasmic mutagenic effects. *Mutat. Res.* 62:239–253 (1979).

10 Feldman, MY. Reaction of nucleic acids and nucleoproteins with formal-dehyde. In *Progress in Nucleic Acid Research and Molecular Biology*, edited by JN Davidson and WE Cohn, vol. 13, pp. 1–44. New York: Academic Press (1973).

11 Boreiko, CJ, DB Couch, and JA Swenberg. Mutagenic and carcinogenic effects of formaldehyde. In *Genotoxic Effects of Airborne Agents*, edited by RR Tice, DL Costa, and KM Schaich, pp. 353–367. New York: Plenum (1982).

12 Kohn, KW. DNA as a target in cancer chemotherapy: Measurement of macromolecular DNA damage produced in mammalian cells by anticancer agents and carcinogens. *Methods Cancer Res.* 16:291–345 (1978).

13 Swenberg, JA. Utilization of the alkaline elution assay as a short-term test for chemical carcinogens. In *Short-Term Tests for Chemical Carcinogens*, edited by HF Stich and RHC San, pp. 48–58. New York: Springer-Verlag (1981).

14 Ross, WE, and N Shipley. Relationship between DNA damage and survival in formaldehyde-treated mouse cells. *Mutat. Res.* 79:277–283 (1980).

15 Ross, WE, DR McMillan, and CF Ross. Comparison of DNA damage by methylmelamines and formaldehyde. *J. Natl. Cancer Inst.* 67:217–221 (1981).

16 Magaña-Schwencke, H, and B Ekert. Biochemical analysis of damage in-duced in yeast by formaldehyde. II. Induction of cross-links between DNA and protein. *Mutat. Res.* 51:11–19 (1978).

17 Chaw, YFM, LE Crane, P Lange, and R Shapiro. Isolation and identification of cross-links from formaldehyde-treated nucleic acids. *Biochemistry*, 5525–5531 (1981).

Part Three

Human Studies

Chapter 13

Human Studies
with Formaldehyde: Introduction

John R. Froines

NIOSH has devoted significant attention to the hazards associated with formaldehyde exposure. The NIOSH Occupational Hazard Survey (NOHS), which was carried out between 1972 and 1974, studied a sample of 5000 separate workplaces and documented formaldehyde use in at least 396 separate industrial settings. The second National Occupational Hazard Survey is just beginning and teams will be going into the field in the very near future to begin to develop further information as to the distribution of chemicals throughout the workplace. It is hoped that the results will confirm our original evaluation that formaldehyde was in use in a wide range of industries and we will have an even better estimate of the percentage of workers actually exposed to formaldehyde.

The original NOHS Survey did confirm our impression that exposure to formaldehyde is common in medical and laboratory environments and in certain parts of the textile, wood, and paper industries. Following the Survey, NIOSH issued a criteria document in 1976, recommending 1 ppm for a 30-min sampling period. The recommendation in the criteria document stated: "Based on reports of irritation, objectionable odor, and disturbed sleep on exposure of formaldehyde at 3/10 ppm and of more general complaints at concentrations greater than

1 ppm, a ceiling value of 1 ppm in the air is proposed as the workplace environmental limit." That recommendation holds today as it did in 1976. The carcinogenic potential of formaldehyde was not known at that time and was not considered in the recommendations. I might say parenthetically that in that criteria document the following recommendation was made: "Epidemiologic investigations of various occupational groups exposed to formaldehyde with data on airborne concentrations associated with clinical findings, if any, would allow refinement of the recommended standard."

More recently, we have been involved in a series of health hazard evaluations at NIOSH in which we investigated acute episodes of illness from exposure to formaldehyde wherein workers complained of irritant sickness such as eye irritation, runny nose, cough, headaches, sore throat, and irritated skin. A recent MMWR article concerned formaldehyde exposure at the embalming laboratory at the Cincinnati College of Mortuary Science. These studies have provided us with estimates of formaldehyde exposure levels in industry. For example:

Industry	Formaldehyde level (ppm)
Fertilizer production	0.2-1.9
Dyestuffs	0.1-5.8
Textile manufacture	0.1-1.4
Resins (non-foundry)	0.1-5.5
Bronze foundry	0.12-8.0
Iron foundry	0.02-18.3
Treated paper	0.14-0.99
Hospital autopsy room	2.2-7.9
Plywood industry	1.0-2.5

We are in the process of developing a current intelligence bulletin on formaldehyde, jointly with OSHA. This document will incorporate the information on the carcinogenicity of formaldehyde that has been developed at this conference and will be released at a later date.

Lastly, NIOSH has been participating in the National Toxicology Program Committee to assess health risks to humans from chronic exposure to formaldehyde. Our participation in that committee and discussions within the Institute have led us to the following conclusions with respect to the epidemiologic research priorities:

1   Carefully designed epidemiologic studies to evaluate the role of formaldehyde in human disease are a high priority at NIOSH. Additional studies of chronic respiratory system disorders that include environmental measures to allow more precise estimates of risk at various exposure levels need to be undertaken.

2　Projects need to be developed within NIOSH and in collaboration with other groups to confirm or deny the association of reproductive effects with exposure to formaldehyde.

3　There are no complete epidemiologic studies to evaluate the carcinogenicity of formaldehyde in human populations at this point. There are several ongoing cohort studies of workers exposed to formaldehyde, e.g., morticians, laboratory workers, and workers in the chemical industry, designed to evaluate the risk of cancer in these groups, but additional research is necessary.

NIOSH hopes to identify additional cohorts for epidemiologic study based on our knowledge of the end uses of the chemical. In this regard, our industry-wide studies branch is monitoring levels of formaldehyde exposure in several industries including wood, paper, the chemical industry, and textiles in order to develop suitable cohorts for mortality studies. Specific mortality studies are being planned in textiles and woodworking.

One of the major difficulties is the development of a proper cohort, and case-control approaches may prove to be more efficacious in elucidating the carcinogenic potential of formaldehyde.

Chapter 14

Controlled Human Studies with Formaldehyde

Ib Andersen

Lars Mølhave

Formaldehyde is an important, very cheap, high volume chemical that is used in a wide variety of products. Exposure to formaldehyde in the occupational environment is very common and is found in a large variety of jobs—in the chemical industry, in plywood and particleboard manufacture, in the textile industry, in the furniture industry, and in hospitals, among many others. An even greater number of people are exposed to formaldehyde in the non-industrial and the home environment due to the widespread use of formaldehyde in building materials such as particleboard, plywood, and insulation materials and in furniture and textiles. In Denmark it is estimated that in 200,000 to 300,000 dwellings, at least 500,000 subjects (approximately 10 percent of the population) are exposed to formaldehyde concentrations above 0.15 mg/cubic meter.

In toxicology, adequate information about the toxicity of a product is rarely available and this applies especially to human exposure studies under controlled conditions. In view of man's almost ubiquitous exposure to formaldehyde and the many complaints about eye and airway irritation, skin problems,

E. Holst performed the statistical analysis.

and other disorders caused by this exposure it is staggering that we are able to report on only five studies on human subjects performed under controlled conditions. More human studies have been performed with mixed exposures and with little or no control of the experimental parameters, e.g., studies in the workplace. Of the five studies under controlled conditions, four (1-4) deal with exposure periods of a few minutes up to 37 min duration; only in one study (5) is the exposure period of 5 h duration.

PREVIOUS STUDIES

The first study under controlled conditions is an English study from 1957 (1), in which 12 healthy males (18-45 years) were exposed in a 100-cubic meter chamber to 17 mg/cubic meter formaldehyde for 30 min. The formaldehyde was generated by bubbling air through a formaldehyde solution, and it was measured by a titration method. When the men entered the chamber, there was considerable nasal and eye irritation, but despite the continued mild lacrimation for some period of time, there was no marked response to the exposure. The eye irritation was not severe and wore off after about 10 min in the chamber.

In a Swedish study by Pettersson and Rehn (2), the odor threshold for formaldehyde was measured on 64 subjects (17-63 years). The subjects were exposed in a climate chamber, the temperature being $22 \pm 1°C$, the relative humidity 40 ± 10 percent, and the air changed 5 times per hour. The air was filtered through charcoal and a high performance particle filter. The air in the chamber was clean and the subjects were exposed to formaldehyde by placing their heads during periods of measurements in a hood supplied by air containing formaldehyde generated by paraformaldehyde powder and then diluted to a given concentration. The formaldehyde concentration was measured by the chromotropic acid method. The exposures were alternately a formaldehyde-air mixture or clean air, the sequence being at random. The task of the subjects was to determine in which of the two exposures formaldehyde was present. Each exposure series was started at concentrations below the subject odor threshold and the concentration was increased in steps until three determinations were correct. The exposure series was repeated six times.

In Fig. 1 the cumulative function of six correct statements during six exposures is shown. It appears that the lowest detectable odor level in this cross-section of the population is about 0.05 mg/cubic meter. The study was an acute exposure study and adaptation to the odor will occur during longer exposures, appearing as an increase of the lowest detectable level. It was found that smokers were significantly less sensitive than were nonsmokers, whereas no differences were observed between subjects of different age or between males and females.

In a Swiss study by Weber-Tschopp et al. (3) healthy students were exposed to formaldehyde in concentrations between 0.04 and 5 mg/cubic meter in order to study the acute irritation effects on airways and eyes. The experiment was

No. of subjects in %

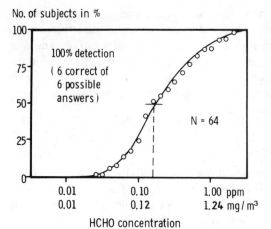

HCHO concentration

Figure 1 The cumulative function of the lowest detectable odor level for formaldehyde in 64 subjects giving six correct answers during six exposures. [*From Pettersson and Rehn (2).*]

performed in a 30-cubic meter climate chamber with an air change rate of 0.1 h. Formaldehyde was generated by continuous injection of 35% formaldehyde solution in a 120°C heated glass tube with a flow through of N_2. The formaldehyde concentration in the chamber was measured continuously by the chromotropic acid method. The standard deviation on the variation of formaldehyde concentrations was 5 percent. The study consisted of two series of experiments. In the first series 33 healthy students in groups of two were exposed continuously to a formaldehyde concentration that increased steadily from 0.04 to 4 mg/cubic meter during 37 min. In the second series 48 healthy students in groups of four were exposed to 0, 1.2, 2.4, 3.6, and 5 mg/cubic meter, the duration of the exposure at each concentration being 1.5 min and intervals between exposures of 8 min duration, when the subjects were in a well-ventilated room without formaldehyde. A questionnaire about eye and airway irritation was answered five times during each experiment and after each questionnaire the frequency of eye blinking was measured.

In Fig. 2 is shown the subjects' impression of the air quality. It appears during both exposures that the air quality is considered worse when the formaldehyde concentration increases. The impression of the subjects was that the air quality during discontinuous exposure was inferior to the air quality during continuous exposure, even if the formaldehyde concentrations were identical. This probably reflects that adaptation takes place during continuous exposure.

The irritation of the eyes and the nose is shown in Fig. 3. In the eyes the continuous exposure was significantly more irritating than the discontinuous exposure, and the opposite was found in the nose. During the continuous exposure the sensitivity of the two organs was about the same. Almost no irritation was reported from the throat. Also the eye blinking frequency was mea-

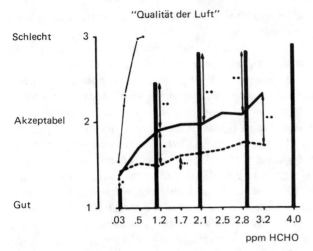

Figure 2 The impression of the air quality given by 33 subjects exposed either continuously during 35 min to formaldehyde (full-drawn line) or to clean air (broken line). The impression from discontinuous exposures to formaldehyde during 1.5-min periods (vertical bars) and from exposure to sidestream cigarette smoke (dotted line) given by 48 subjects is also shown. The levels of significance are given as (•), •, and ••, being $p < 0.05$, < 0.025, and < 0.01, respectively. [*From Weber-Tschopp et al. (3).*]

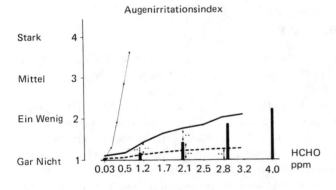

Figure 3 The irritating effects of formaldehyde on eyes and nose (upper and lower half, respectively). Explanation as for Fig. 2. [*From Weber-Tschopp et al. (3).*]

sured. At 2.4 mg/cubic meter and above, the frequency was significantly higher than during the control condition.

In the English thesis "Human reflex bronchoconstriction as an adjunct to conjunctival sensitivity in defining the threshold limit values of irritant gases and vapours" (4), a few experiments with formaldehyde exposure were performed. The formaldehyde-air mixture was obtained by bubbling air through Formalin solution, and the concentration was measured by a colorimetric method. The subjects inhaled the air through a mouth tube and air for exposure of the eyes was provided to a pair of goggles. Four subjects were exposed to 10 mg/cubic meter and 16 mg/cubic meter. All the subjects except one showed a drop in conductance ranging from 19 to 43 percent at the lower concentration and ranging from 50 to 108 percent at the higher concentration. At the low concentration the subjects reported an irritation in the throat after two to five breaths and similar reports were given but worse irritation stated at 16 mg/cubic meter. During eye exposures, 5 of 6 subjects exposed to 16 mg/cubic meter had symptoms after 4–15 s, whereas 4 of 5 exposed to 11 mg/cubic meter had no eye irritation. Apparently the lower airways were more sensitive to formaldehyde than the eyes, but it has to be borne in mind that the air was inhaled through a mouth tube and therefore by-passed the nose, thus eliminating reactions elicited by this organ.

A FIVE-H EXPOSURE STUDY

The only study in which subjects have been exposed during a period of several hours was performed in the climate chamber at the Institute of Hygiene, Århus, Denmark (6). Some preliminary results have been reported earlier (5).

The subjects were 16 healthy students, 5 females and 11 males, average age 23 years and age range 20 to 33. Five were smokers, but only one was a heavy smoker, i.e., 20 cigarettes a day. None of them had been or were exposed to formaldehyde, and all had apparently healthy upper airways. They were all habitually nasal breathers and nobody had a history of chronic or recent acute respiratory disease. They were studied in groups of four, each group undergoing four different exposures on four consecutive days. The exposures were 0.3, 0.5, 1.0, and 2.0 mg formaldehyde per cubic meter of air, the order being arranged in a latin square design.

Each day began with a control period of 2 h duration with exposure to clean air at $23 \pm 0.5°C$, 50 ± 5 percent relative humidity, air velocity 10 ± 3 cm/s, and air supply rate of 500 cubic meter/h, the air being fresh outdoor air filtered through absolute filters and charcoal filters.

The measurements each day during the control period together with a set of measurements performed prior to the first day (day 0) were taken as control values. The thermal and atmospheric conditions were maintained during the whole experiment, but after the control period, formaldehyde was added to the air. After about 1 h a steady state concentration was reached and this was maintained during the rest of the day.

The formaldehyde was generated by passing air through an oven in which there was a container with paraformaldehyde heated to about 80°C. The paraformaldehyde had previously been baked at 120°C for 5 h to stabilize its degassing of formaldehyde. The weight loss of the paraformaldehyde was measured continuously and was kept constant by controlling the temperature of the oven. The formaldehyde concentration in the chamber air was measured by collection of 1-h air samples and analysis by the chromotropic acid method. The variation was within ±20 percent from the target values.

During each day three identical series of measurements on subjects took place. The first series was performed during the control condition, the second after 2–3 h exposure and third after 4–5 h exposure.

During each series of measurements we carried out the following: We measured nasal mucociliary flow by external detection of the motion of a resin particle (diam. 0.056 mm) labeled with 2 μc ^{99m}Tc placed under direct vision on the superior surface of the inferior turbinate, which is in the mainline of inspiratory airflow. This technique is fully described in (7). After the completion of that test we measured nasal airflow resistance through an oro-nasal plastic mask attached to a penumotachometer. Also the forced expiratory vital capacity (FVC), $FEV_{1.0}$, $FEF_{25-75\%}$, and the odor threshold for ethyl valeriate were measured.

Throughout each day we asked the subjects to adjust the pointer on a voting machine expressing the degree of airway irritation on a scale of 1 for complete comfort and 100 for severe discomfort. The position of each individual's pointer was concealed from the other subjects and it was recorded continuously. At the end of each day and the following morning the subjects were questioned as to the degree and nature of discomfort they had experienced.

Fluids were permitted ad lib, lunch was served during the third hour, and no smoking was allowed.

The performance of the subjects was measured by tests of 30 min duration three times a day—once in the control period in clean air and twice during the exposure period. The three tests were numerical addition, multiplication, and card punching, each of 15 min duration. The addition test was performed in each period, the multiplication test in the first exposure period only, and card punching in the control period and second exposure period. The tests have been described earlier (8).

All data were first analyzed by nonparametric tests; then an analysis on the differences between the individual observations from run 1 and runs 2 and 3 was performed. To obtain a suitable agreement between the data and the assumptions of the analysis of variance (ANOVA), a data transformation was required in some calculations and square root transformations proved to be sufficient in most cases. The calculations were performed on a UNIVAC 1100/82 computer using the GENSTAT ANOVA macro. Our level of significance is 5 percent.

RESULTS

Nasal Mucus Flow

The mean flow rate at the preliminary investigation of the subjects (day 0) in clean air and after 4–5 h exposure to four different concentrations of formaldehyde is shown in Fig. 4 for five different areas of the nose; slits 1–2 and slits 5–6 being the most anterior and posterior parts of the ciliated portion of the nose, respectively. It appears that the mucus flow rate slows down at higher formaldehyde concentrations, but no further deceleration is found above 0.5 mg/cubic meter. There is a significant decrease in the mucus flow rate in the anterior two thirds of the nose (slits 1–4), whereas no significant difference is found at slits 4–6. The analysis on the differences showed significant differences only at slits 1–2.

In Fig. 5 is shown the average mucus flow in the middle third of the nose at exposure to four different formaldehyde concentrations. It appears that there is no difference between the reduction after 1–3 and after 4–5 h exposure. A similar pattern was found in the anterior third of the nose, whereas the measurements in the posterior part of the nose were not different during clean air and during formaldehyde exposures.

Airway Resistance

The results of the flow measurements in the airways are shown in Fig. 6. The average standard deviations from the mean for the nasal pressure drop, VC,

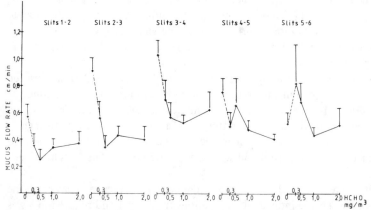

Figure 4 The average mucus flow rate in clean air (0) and after 4–5 h exposure to formaldehyde at four concentrations. Slits 1–3, 3–4, and 4–6 represent the anterior, middle, and posterior thirds of the ciliated part of the nose, respectively. One standard deviation of the mean is shown as a vertical bar.

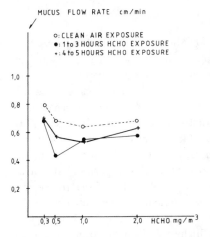

MUCUS FLOW RATE cm/min

o: CLEAN AIR EXPOSURE
●: 1 to 3 HOURS HCHO EXPOSURE
+: 4 to 5 HOURS HCHO EXPOSURE

Figure 5 The average mucus flow rate in the middle third of the nose (slits 3–4) at four formaldehyde concentrations. The three curves represent exposures to clean air (0) and to formaldehyde, the latter being divided into 1–3 (●) and 4–5 h exposure (+). The average standard deviation of the mean was 0.11 cm/min.

$FEF_{25-75\%}$, and $FEV_{1.0}$ are 5, 0.28, 0.28 and 0.24, respectively. No significant changes were found in the rhinomanometric parameters or in the other airway resistance parameters.

Odor Threshold

After 1–3 and 4–5 h exposure to formaldehyde (2 mg/cubic meter) there was a significant increase in the odor threshold for ethyl valeriate; at no other concentration did significant threshold changes occur.

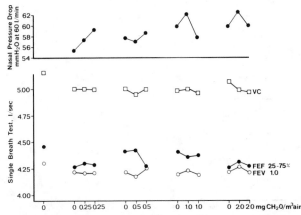

Figure 6 The variation with time of the averages of nasal pressure drop (upper part), of vital capacity (VC), forced expiratory flow ($FEF_{25-75\%}$), and forced expiratory volume during the first second of the expiration ($FEV_{1.0}$) (lower part). The interval between the measurements each day is approximately 2 h.

Discomfort

The average subjective discomfort ratings are shown in Fig. 7. It appears that during the first 2 h of the exposure period no discomfort is caused by 0.3 or 0.5 mg/cubic meter. In the remaining part of the exposure period these concentrations caused increasing discomfort, 0.3 mg/cubic meter being more uncomfortable than 0.5 mg/cubic meter. At the two higher concentrations discomfort is reported already during the first hour's exposure. The discomfort increases during the following 2 h and then stabilizes at the 1.0 mg/cubic meter exposure, whereas the discomfort then decreases at the 2.0 mg/cubic meter exposure, indicating that acclimatization occurs at this exposure but not at lower concentrations. After $2\frac{1}{2}$ h exposure, 13, 14, 9, and 6 of the 16 subjects had no discomfort at 0.3, 0.5, 1.0, and 2.0 mg/cubic meter, respectively. At the end of the day the number of subjects without discomfort were 7, 13, 10, and 6, respectively. At these concentrations the highest individual votings recorded were 30, 20, 40, and 50 scale units, respectively, but even at the highest concentration the subjects' average discomfort never exceeded 18 scale units, which is in the middle of the "slight discomfort" range.

After the exposure the subjects were asked about the character of their symptoms. The complaints were mainly conjunctival irritation and dryness in the nose and throat. After the exposures to 0.3, 0.5, 1.0, and 2.0 mg/cubic meter, 3, 5, 15, and 15 subjects, respectively, had these complaints. There were no carryover symptoms; the following morning all subjects were without complaints.

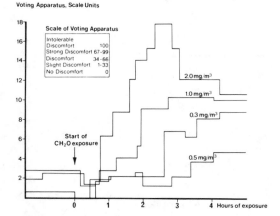

Voting Apparatus, Scale Units

Figure 7 Variation with time of the mean discomfort vote in scale units at four different concentrations of formaldehyde. In the control period, clean air was supplied to the subjects.

Performance

The performance of the subjects measured by speed and accuracy in addition, multiplication, and transfer of numbers to punchcards was the same at all conditions.

DISCUSSION

Formaldehyde is very soluble in water (980 1 per liter water at $20°C$ and 760 mm Hg) and has a high chemical reactivity. These properties explain why formaldehyde irritates, especially the mucus membranes of the eyes and the upper airways. During nasal respiration the formaldehyde like sulfur dioxide (7) (which has a solubility of 38 1 per liter water) will be absorbed in the anterior part of the nose, for which reason only high formaldehyde concentrations should be able to affect the lower airways. The observation that mucus flow decreases in the anterior two thirds of the ciliated epithelium during formaldehyde exposure whereas no effect is seen in the posterior third (Fig. 4) also indicates that the absorption mainly takes place in the anterior part of the nose. There was no difference between the values obtained after 1-3 h formaldehyde exposure and those obtained after 4-5 h exposure, so no saturation effect occurred during this period.

A comparison between the effects of SO_2 and formaldehyde on the airway shows that big differences exist. In the same molar concentration, 2 mg formaldehyde/cubic meter and 5 mg sulfur dioxide/cubic meter (2 ppm), the former has no effect on the nasal and tracheobronchial resistance whereas the latter causes a strong increase in both (7). The decelerating effects on nasal mucus flow are of the same magnitude for the two substances, whereas the discomfort due to formaldehyde is less than that caused by SO_2.

The defense against airborne noxious substances is, at least in part, dependent on a rapid removal of impacted or absorbed material by the mucociliary route to the pharynx, where the mucus is swallowed. Any impairment in the speed of this clearance may be accompanied by increased susceptibility to infection.

The present threshold limit value (TLV) for formaldehyde in the United States is 3.6 mg/cubic meter. In the preface of the TLV booklet (9) it is stated that "Because of the wide variations in individual susceptibility, however, a small percentage of workers may experience discomfort from some substances at concentrations at or below the TLV." The experience from our 5-h exposure study at 2 mg/cubic meter, wherein 15 of 16 subjects complained of irritation of conjunctiva, nose, and throat after this exposure and physiological reactions such as decrease in nasal mucus flow and in odor perception were found, as well as the experiences from the other studies at controlled conditions, especially Pettersson and Rehn (2) and Weber-Tschopp et al. (3), make us believe that the TLV value for this substance should be lower than 3.6 mg/cubic meter.

These observations and the possibility of developing hypersensitivity to formaldehyde during occupational exposure to this substance has led to a decrease in the Danish TLV value to 0.40 mg/cubic meter beginning in 1983 (10). At present the Danish TLV value and the German MAK value are 1.2 mg/cubic meter.

CONCLUSIONS

The conclusions to be drawn from the five controlled human studies with formaldehyde reviewed may be described in three main categories—the effect of formaldehyde on physiological parameters, on subjective discomfort, and on performance.

Among the physiological parameters, no effects on nasal resistance, expiratory vital capacity, and tracheobronchial resistance have been found in our study exposing subjects during 5 h to formaldehyde in concentrations from 0.3 to 2.0 mg/cubic meter. At mouth breathing a drop in conductance occurs after few breaths at 10 mg/cubic meter (4).

The nasal mucus flow rate in the nose decreases during exposure to formaldehyde, but the effect, probably due to the high solubility of formaldehyde in water, is found only in the anterior two thirds of the ciliated part of the nose, and the response does not increase at concentrations above 0.5 mg/cubic meter or at prolongation of the exposure period from 3–5 h.

The lowest detectable odor level during an acute formaldehyde exposure of a cross-section of the population is about 0.05 mg/cubic meter, and 50 percent of the population will always detect about 0.21 mg/cubic meter. During longer exposure periods an adaptation to the formaldehyde odor occurs. Smokers are less sensitive than nonsmokers, but age and sex do not influence the sensitivity (2). Exposure to formaldehyde, 2 mg/cubic meter, during 1–5 h increases the odor threshold for ethyl valeriate, whereas no effect is found at lower concentrations. The subjects' impression of the quality of air is that the air quality deteriorates when the formaldehyde concentration increases. This effect is found at 1.2 mg/cubic meter and above and is more pronounced during discontinuous than during continuous exposure, probably reflecting an adaptation during the latter method of exposure. The frequency of eye blinking is increased at exposure to 2.1 mg/cubic meter and above (3).

Subjective discomfort caused by 30-min continuous exposure to formaldehyde up to 3.6 mg/cubic meter is slight irritation of the eyes and of the nose (3), whereas irritation of the throat is experienced only after longer exposure periods. The irritation is directly proportional to the formaldehyde concentration. During discontinuous exposure the irritation of the eye is even smaller (3), whereas it is greater in the nose, indicating that nasal adaptation takes place. At very short eye exposures of about 15 s duration, eye irritation is experienced from 11 mg/cubic meter and above (4). During exposures of 5 h duration at formaldehyde concentrations of 0.3 mg/cubic meter irritation is reported by 3 of 16 subjects and the number of subjects affected increases directly proportional with the concentration. After 15 h in clean air, no irritation persists. These irritation effects and the fact that 6 of 16 subjects had no complaints after 5 h exposure to 2.0 mg/cubic meter illustrate the wide variations in individual susceptibility.

Formaldehyde exposures during 5 h to concentrations up to 2.0 mg/cubic meter do not affect the subject's ability to perform mathematical tests.

REFERENCES

1 Sim, VM, and RE Pattle. Effect of possible smog irritants on human subjects. *JAMA* 165:1908–1913 (1957).

2 Pettersson, S, and T Rehn. Determination of the odor threshold for formaldehyde. *Hygien & Miljö* No. 10 (1977).

3 Weber-Tschopp, A, T Fischer, and E Grandjean. Irritating effects of formaldehyde on men. *Int. Arch. Occup. Environ. Health* 39:207–218 (1977).

4 Douglas, RB. Human reflex bronchoconstriction as an adjunct to conjunctival sensitivity in defining the threshold limit values of irritant gases and vapours, thesis. London: TUC Centenary Institute of Occupational Health, London School of Hygiene and Tropical Medicine (1974).

5 Andersen, I. Formaldehyde in the indoor environment–health implications and the setting of standards. In *Indoor Climate*, edited by PO Fanger, and O Valbjørn, pp. 65–77. Copenhagen: Danish Building Research Institute (1979).

6 Andersen, I, and GRL Lundqvist. Design and performance of an environmental chamber. *Int. J. Biometeor.* 14:402–405 (1970).

7 Andersen, I, GR Lundqvist, PL Jensen, and DF Proctor. Human response to controlled levels of SO_2. *Arch. Environ. Health* 28:31–39 (1974).

8 Andersen, I, PL Jensen, P Junker, A Thomsen, and D Wyon. The effects of moderate heat stress on patients with ischemic heart disease. *Scand. J. Work. Environ. Health* 4:256–272 (1976).

9 TLVs for chemical substances in workroom air. Cincinnati, Ohio: American Conference of Governmental Hygienists (1980).

10 Hygiejniske graensevardier. Copenhagen: Arbejdstilsynet (1980).

Chapter 15

Formaldehyde: Effects on Animal and Human Skin

Howard Maibach

Formaldehyde has daily contact with the skin of most people. To fully characterize its many effects is complex: this chapter intends to summarize current experience and areas requiring further effort. The approach taken is historical and profile (1), and the many cutaneous effects of formaldehyde are presented.

PERCUTANEOUS PENETRATION

Formaldehyde penetrates human and animal skin. If some amount of the molecule or its metabolites did not penetrate, allergic contact dermatitis could not occur.

Percutaneous penetration is complex; ten separate steps have been identified (2), several of which can be quantitated. We hope to detail these steps for formaldehyde.

An aspect of percutaneous penetration of formaldehyde that will perhaps be the most complex is its distribution and metabolism within the skin. Because formaldehyde is highly reactive, one must assume that degradation occurs quickly.

Once these quantitative data become available, the dermatotoxicologist will be able to make more meaningful extrapolations (rather than all or none estimates) of hazard in each given use situation. Critical questions will be how much penetrates at a given anatomic site under specified use conditions and how much will adhere to the skin (substantivity or persistence) or will formaldehyde (e.g., in a shampoo or detergent) deposit on skin and penetrate in significant amounts. Other aspects of the problem are outlined elsewhere (2).

CUTANEOUS IRRITATION

Anyone working with Formalin is aware of the effects on the fingertips. Oddly enough, although its irritancy potential is generally acknowledged, little quantitative data exist. In humans, the 21-day cumulative irritancy assay is a standard method for ranking the cumulative irritancy potential of a chemical (or final formulation) compared to a standard reference compound. No such data exist for Formalin.

The only Formalin relative irritancy data we are aware of were sponsored by the CPSC. The test system was the OET (open epicutaneous test). In this assay, the first several open applications allow one to make quantitative estimates of irritancy potential (3). In this assay, marked irritation occurred at 30 and at 10 percent Formalin. Even 3, 0.1, and 0.03 percent produced irritation (Table 1).

This is an estimate of irritancy that, although reliable for the guinea pig, cannot be easily extrapolated to humans because a reference marker compound was not employed. Similar quantitative data are needed for humans. Nevertheless, it is likely that Formalin or final formulations containing Formalin at 300 ppm or greater may, depending on the actual use (anatomic site, wash off versus not wash off, and frequency of dosing), induce clinical irritation. Clearly, more toxicologic data should be generated in this important area.

Table 1   Guinea Pig Irritancy Scores at Day 7 in the Open Epicutaneous Test (OET)

Test concentration (%)	Scores in individual animals[*]
30	4, 4, 4, 4, 3, 3, 4
10	3, 3, 2, 2, 2, 1, 1/2
3	0, 0, 1/2, 1/2, 1/2, 1/2, 1/2, 1/2, 1/2
1	0, 0, 0, 0, 0
0.3	0, 0, 0, 0, 0
0.1	1, 0, 0, 0, 2, 0
0.03	0, 0, 1, 1, 0, 0

[*]Scoring system: 0 = none; 1/2 = very slight erythema; 1 = slight erythema; 2 = moderate erythema and edema; 3 = strong erythema and edema; 4 = very strong erythema and edema.

PHOTOIRRITATION (PHOTOTOXICITY)

Some patients with allergic contact dermatitis to Formalin have the dermatitis in the exposed area. These are also the areas of sunlight exposure. There are no reports of the photoirritation (phototoxic) potential of Formalin. Unfortunately, previous experience has taught us that the prepared mind identifies a clinical problem. With the ubiquitous nature of Formalin, it appears prudent to obtain the information as to whether this material has phototoxic potential.

This assay can easily be performed in several animal species and humans. The techniques for these assays are reviewed elsewhere (4).

ALLERGIC CONTACT DERMATITIS

Formalin produces allergic contact dermatitis. This is an area in which there is considerable emotional impact; some people desire to ban Formalin because of its cutaneous sensitization potential. In certain countries there is extensive prohibition of its use for this reason.

Allergic contact dermatitis is no different than any other form of toxicity: it requires reliable quantitative data and judgment in interpretation. Several aspects of the issue are complicated but may be better understood if each of its parts is dissected.

Making the Diagnosis of Allergic Contact Dermatitis

As noted above (under irritation) Formalin is an irritant. It is even more irritating when applied to the skin under an occlusive patch, as is done in the diagnosis of allergic contact dermatitis (the diagnostic patch test).

Until recently, the standard concentration for this patch testing was 5% Formalin in water. Unfortunately, as emphasized by Epstein, this produces irritancy in a significant number of nonsensitized (virgin) controls (5).

Patch testing with materials of low irritancy potential, i.e., a large separation between the concentration that elicits an allergic response and an irritant response, permits testing by the noviate with little experience. When testing with Formalin, this is not the case; minimal reactions (usually irritation) are generally regarded as nonallergenic by the experienced dermatoallergist. Unfortunately, much patch testing has been performed by physicians not adequately trained in contact dermatitis; hence a significant number of irritant responses have been incorrectly interpreted as allergic (5).

The current standard is 2% in water—when the diagnostic dermatologist utilizes the Al Test patch (Astra, Sweden). Even with this concentration, Formalin remains close to the irritating level, as determined in control subjects.

For this reason, one must be careful of false-positive responses. When patch testing with materials close to the irritation level on patch testing, it is necessary to confirm the reaction, technically and clinically. This is especially important when battery testing is performed; the battery refers to an application of multiple patch tests to a patient in an attempt to identify an allergen as the cause of an allergic dermatitis.

To confirm the reaction, it is customary to repeat the patch test—with only the Formalin, as a single patch, being applied. *If this is positive, it is still necessary to carefully correlate the clinical history and, if the circumstances permit, performing a provocative use test.* This refers to the open application in use rather than a patch test of the Formalin-containing material. An in-depth review of the problems of identifying a false-positive patch test is presented elsewhere (6).

Recently, a new patch test chamber has become popular in much of the western world. With this unit, the Finn Chamber (Epitest), the concentration of Formalin should be 1 percent in water, instead of the 2 percent employed with the Al Test (Astra, Sweden).

Cross-Reactions

It is of great practical importance to know whether a given patient will cross-react to related aldehydes. Specific information is available with glutaraldehyde; cross-reaction has not been identified (7).

Induction Concentration (to Produce Sensitization) and Elicitation Concentration

In the guinea pig, utilizing the OET method, Formalin will induce sensitization with ease. We studied several concentrations: 30, 10, 3, and 0.3 percent. All sensitized (Table 2). Only by reducing the induction concentration to 0.03 percent did we fail to sensitize.

In human experiments, we lack information as to the threshold (as noted above) for sensitization. That such a threshold must exist is clear; otherwise, a much larger number of our population would be sensitized from their daily formaldehyde exposure.

Table 2 Open Epicutaneous Test in Guinea Pigs

Induction concentration (%)	Number sensitized challenge at 1%
30	2/7
10	5/8
3	3/8
1	2/6
0.3	2/6
0.1	0/6

In one human experiment, utilizing the modified Draize repeat insult patch test method, we utilized an induction concentration of 10 or 5 percent (in water), challenging at 1 percent (8). We sensitized 8 percent of the 154 subjects exposed. A 0.01 percent concentration elicited a reaction in one of the five subsequently tested with further dilutions.

Attempting to quantitate the relevance of these data, formaldehyde-containing cosmetics were applied topically to the sensitized subjects: 5 of 10 reacted to a skin lotion containing 0.5 percent formaldehyde; 4 of 10 reacted to a cream rinse containing 0.4 percent formaldehyde; and 2 of 10 reacted to a bubble bath containing 0.6 percent formaldehyde.

In the human maximization test, Kligman exposed 25 subjects to 5 percent Formalin and challenged with 1 percent (9). Of the 25 subjects, 18 (72 percent) were judged sensitized. The above data are minimal; additional experience is needed in animals and in humans to determine the threshold for the induction of sensitization. This information should help in setting exposure limits in occupational and nonoccupational usage.

Equally important to the determination of induction concentration required for sensitization is the elicitation data: once one is sensitized, what is the lowest level that will produce dermatitis?

Horsfall determined this concentration to be less than 1 ppm by interdermal injection (10). He also utilized an artificial use test (a 40-min soak) and noted a similar result. This work requires contemporary verification.

Jordan et al. noted in patch testing that four of nine sensitized subjects reacted at approximately the 30-ppm level (11).

More information is needed in this area. Once sensitized, the subject is in a more vulnerable position in terms of developing dermatitis to a lower exposure level.

Provocative Use Test

It is important to appreciate that most of the above data (except for the guinea pig) refers to that obtained with an occlusive patch test. This increases the sensitivity of the method, presumably by enhancing percutaneous penetration. It is essential to know the threshold level to elicit ordinary usage without a patch test. The usage test (or the provocative use test) consists of the open application (without occlusion) over a period of a week or more.

Jordan et al. applied Formalin-containing material to the axilla: 2 of 11 subjects reacted down to 30 ppm. We made similar observations with a Formalin-containing antiperspirant: 7 of 14 patch test-positive subjects reacted at 390 ppm, and 2 of 14 reacted at 150 ppm. The threshold for no response was 80 ppm (Maibach H, and T. Franz, Unpublished data).

These are small test samples; presumably the level for reactivity would be even lower in a larger population.

Again, it is important to broaden the experimental base for the elicitation exposure in the provocative use test. This critical information will allow us to set "no effect" levels even in sensitized subjects. These levels will be effected by use concentrations: vehicle, anatomic site, wash off versus leave on, and frequency of application.

Frequency of Allergic Contact Dermatitis (Epidemiology)

Most of our experience in this field has been in determining the number of patch test responders seen in dermatologic testing centers. In recent years, significant amounts of this information have become available. The International Contact Dermatitis Research Group found a 3.5 percent reactivity in 4825 patients, whereas in Poland, Rudzki et al. had a 6.3 percent rate in 1205 patients (12, 13). In North America, the North American Contact Dermatitis Research Group observed a 4 percent rate in 1000 patients (14).

It is important to consider this information in context. These are uncorrected patch test responderer data. No attempt was made to correct for the excited skin state ("angry back") or to determine clinical relevance. Until such a study is done, we assume that these numbers of responders in eczema patients are inflated. In general, the number of nonreproduceable patch tests (the excited skin state) must be at least 40 percent (5). It may even be higher on this type of marginally irritating material.

Frequency in Normals

Unfortunately, we have no experience with this type of epidemiologic investigation in normal (non-eczema) subjects. We have data for several other materials as a frame of reference. For instance, in San Francisco, approximately 10 percent of all females are allergic to nickel and approximately 1 of every 80 males and females is allergic to neomycin (15). It is critical that we obtain similar data for Formalin. This is a most relevant question in determining risk-benefit analyses.

HISTORY OF FORMALIN DERMATITIS: EXPOSURES

There are numerous exposures to formaldehyde. This includes the formaldehyde resins utilized in increased area resistant textiles, nitrogenous fertilizers, disinfectants, preservatives, deodorizers, paper manufacturer, medicaments, pathology, and the rubber industry. An extensive review of this literature is presented elsewhere (16).

FACTORS AFFECTING WHETHER A SUBJECT
WILL DEVELOP DERMATITIS FROM FORMALIN

Extrapolating from toxicologic sensitization assays to human use is difficult. A critical example of this is the preserving of formaldehyde of shampoos with Formalin. This is almost ubiquitous in the United States. Nevertheless, with this application, clinical disease caused by Formalin is almost unknown. Presumably, the mechanism is that the Formalin is not substantive enough to stay on the skin and penetrate from such an application.

Many of the other aspects of this must be worked out in the years to come. One key factor will be regional variation in percutaneous penetration. A review of this is presented elsewhere (2).

PHOTOALLERGY

As mentioned above, some Formalin dermatitis (with the exception of clothing) occurs in exposed areas. Is there a photo element to this allergy? We do not know: there are no published assays on the photoallergic potential with Formalin.

We believe it prudent that an appropriate toxicologic assay (such as the photoDraize technique) be performed.

CONTACT URTICARIA

The contact urticaria syndrome refers to an old phenomena that has recently become popularized. The syndrome consists of at least three separate types: nonimmunologic, immunologic, and of uncertain mechanisms.

Since this syndrome has been looked for only recently, our information is fragmentary at best. For instance, there has been no attempt to do a Formalin toxicologic assay for contact urticaria potential.

Helander has already documented a case of contact urticaria from Formalin-treated leather (17). The urticaria occurred not only at the site of exposure but also at distant sites. It was not proved but is probable that this was type B (immunologic).

A current review of the contact urticaria syndrome is found elsewhere (18).

DERMATITIS FROM INHALATION

Our knowledge in this area is also fragmentary. An important observation is that of Horsfall (10), who documented that the inhalation of 10 ppm in sensitized subjects would produce vesicles. Additional information is needed in this regard.

CHEMICAL LEUKODERMA

Vitiligo is a disease of unknown etiology. It is mimicked by a chemically induced erruption that can be morphologically indistinguishable.

It is known that certain phenols and catechols will produce depigmentation. Furthermore, P. tertiary butyl phenol formaldehyde will also produce this type of depigmentation.

Unfortunately, again there are no toxicologic assays for the ability of formaldehyde to produce depigmentation (mimicking vitiligo).

Because the assay is so straightforward, this should be done. Reference 19 details this assay and its background.

CONCLUSION

Formaldehyde has many cutaneous effects. We have learned to live with many of these—in an emperic fashion. It is likely that we could learn to do an even better job in setting appropriate limits, methods of application, and vehicles if we obtained additional toxicologic studies of a quantitative nature. Some aspects of this profile have been noted above. We are optimistic that when these additional data are obtained, we will be able to handle Formalin in a safer fashion.

REFERENCES

1 Maibach, H, and F. Marzulli. Toxicologic perspectives of chemicals commonly applied to the skin. In *Cutaneous Toxicology*, edited by V Drill, pp. 247–264. New York: Academic Press (1977).

2 Maibach, H, and R Wester. Ten steps in percutaneous penetration. *Drug Metab. Rev.* (in press).

3 Klecak, G. Identification of contact allergens: Predictive tests in animals. In *Dermatotoxicology and Pharmacology*, edited by F Marzulli and H Maibach, pp. 305–340. Washington, D.C.: Hemisphere (1977).

4 Maibach, H, and F. Marzulli. Phototoxicity (photoirritation) of topical and systemic agents. In *Dermatotoxicology and Pharmacology*, edited by F Marzulli and H Maibach, pp. 211–224. Washington, D.C.: Hemisphere (1977).

5 Epstein, X, and H Maibach. Formaldehyde allergy: Incidence and patch test problems. *Arch. Dermatol.* 94:186–190 (1966).

6 Maibach, H. The excited skin state. In *Allergy*, edited by J Ring. Munich: Springer Verlag (in press).

7 Maibach, H. Glutaraldehyde: cross-reactions to formaldehyde? *Contact Derm.* 1:326–327 (1975).

8 Marzulli, F, and H Maibach. Antimicrobials: Experimental contact sensitization in man. *J. Soc. Cosmet. Chem.* 24:399–421 (1974).

9   Magnusson, B, and A Kligman. *Allergic Contact Dermatitis in the Guinea Pig.* Springfield, IL: A. Thomas (1970).
10 Horsfall, F. Formaldehyde hypersensitivitiveness: An experimental study. *J. Immunol.* 27:569–581 (1934).
11 Jordan, W, W Sherman, and S King. Threshold responses in formaldehyde-sensitive subject. *J. Am. Acad. Dermatol.* 1:44–48 (1979).
12 Fregert, S, N Hjorth, B Magnusson, et al. Epidemiology of contact dermatitis. *Trans. St. John's Hosp. Dermatol. Soc.* 55:17 (1969).
13 Rudzki, E, and D Kleniewska. The epidemiology of contact dermatitis in Poland. *Br. J. Dermatol.* 83:543 (1970).
14 North American Contact Dermatitis Group: Epidemiology of contact dermatitis in N. American: 1972. 108:537 (1973).
15 Prystowsky, S, A Allen, R Smith, et al. Contact sensitivity to nickel, neomycin, ethylenediamine, and benzocaine in a general population. *Clin. Res.* 26:301 (1978).
16 Cronin, E. *Contact Dermatitis.* London: Churchill Livingstone (1980).
17 Helander, I. Contact urticaria from leather containing formaldehyde. *Dermatologica* 113:1443 (1977).
18 Von Krogh, G, and H Maibach. The contact urticaria syndrome. *J. Am. Acad. Dermatol.* (in press).
19 Gellin, G, H Maibach, M Misiaszek, et al. Detection of environmental depigmenting substances. *Contact Derm.* 5:201–213 (1979).

ADDENDUM

Since preparation of this chapter two texts with extensive coverage on Formalin-skin effects have been published: Foussereau, J et al.: *Occupational Contact Dermatitis*, Copenhagen: Munksgaard (1982); and Maibach, H and Gellin, G (eds.): *Occupational and Industrial Dermatology*, Chicago: Year Book (1982).

Chapter 16

Effects of Formaldehyde on the Respiratory System

John Gamble

Formaldehyde potentially affects the respiratory system in three ways: irritation, airflow obstruction or asthma or both, and cancer. Mucous membrane or upper respiratory tract irritation is the most commonly reported effect. Eye, nose, and throat irritation are the most sensitive symptoms. Asthma has been reported but there are few confirmed cases showing specific responses to formaldehyde. The evidence for airflow obstruction is not convincing. There is evidence from animal studies that formaldehyde causes nasal cancer; however, there are no adequate epidemiological studies to assess the risk of respiratory cancer to humans exposed to formaldehyde.

The relevant literature on controlled human exposure, case reports, epidemiological studies, and animal studies relating to formaldehyde exposure will be summarized. Formaldehyde is a widely used industrial chemical. Its greatest use is in the manufacture of urea-, phenol-, acetol-, and melamine-formaldehyde resins, but it has a variety of other uses: agriculture, paper, textile and dyestuff manufacture, medicine, cosmetics and deodorants, disinfectants and fumigants, embalming, concrete and plaster, leather, photography. Both the makers and consumers of formaldehyde products are potentially exposed. Despite the extensive use of formaldehyde, only irritant effects are adequately documented. The effects of formaldehyde on the airways are not conclusive; the prevalence of occupational asthma in exposed populations is unknown, and although exposure levels are not well documented, they appear to be high. No satisfactory mortality study of a formaldehyde-exposed population has been reported.

IRRITATION

An irritant produces inflammation of surface tissues such as eye and membranes of the respiratory tract. While the inflammatory process is similar in all tissues, the symptoms will differ depending on the site. Pulmonary edema and pneumonia are the result of lung inflammation, bronchitis and bronchopneumonia of bronchi inflammation, and rhinitis, pharyngitis, and laryngitis of upper airway inflammation. Formaldehyde is highly soluble in water, so its site of action (in the absence of a particular carrier) is primarily in the upper respiratory tract.

Controlled Studies

Sim and Pattle (1) exposed 12 men for 30 min to 13.8 ppm formaldehyde. The exposure resulted in mild lacrimation and considerable eye and nasal irritation. The eye irritation subsided after about 10 min of exposure.

Schuck and co-workers (2) exposed 12 subjects for 5 min to low levels of formaldehyde and recorded eye irritation and the rate of eye blinking. "Pure air" containing $\leqslant 0.02$ ppm formaldehyde yielded light eye irritation (4 on a scale of 24; a score greater than 16, rated as severe eye irritation, caused lacrimation in more than 50 percent of the subjects). The average eye irritation to formaldehyde at 0.3 ppm, 0.5 ppm, and 0.8 ppm was light (4), medium (7), and severe (12) irritation, respectively. Below 0.3 ppm there was no dose-response relation (that is, the irritation was slightly higher at concentrations less than 0.3 ppm than at 0.3 ppm). Subjects were able to detect formaldehyde at concentrations as low as 0.01 ppm. Eye blinking rate tended to increase as eye irritation decreased and acted to reduce irritation.

Weber-Tschopp et al. (3) exposed 33 subjects for 35 min to 0.03–3.2 ppm formaldehyde and measured eye, nose, and throat irritation, odor, "desire to leave the room," and eye blinking rate. The average response at 0.03 ppm was the same as the control group, with an approximately linear increase in average response above 0.03 ppm formaldehyde. Significant differences between exposed and controls occurred at 1.2 ppm for eye and nose irritation and "desire to leave the room," at 1.7 ppm for eye blinking rate, and 2.1 for throat irritation. Adaptation appeared to occur, as exposure for only 1.5 min at the same concentrations resulted in higher average responses.

Rader (4) exposed two groups of 5 male medical students for 1 h to 0, 0.1, and 0.5 ppm (group 1) and 1, 3, and 5 ppm formaldehyde (group 2). The most frequent complaints in order of frequency were odor, eye irritation, and irritation of the nose and throat region. Lacrimation was about half as frequent as the first group of symptoms. Nasal secretion and perception of dehydration were the least frequent (about 25 percent that of eye and nose symptoms). At 0.5 ppm formaldehyde all complaints except increased nasal secretion were more frequent in air at a lower relative humidity (40 percent) than at higher relative humidity (88 percent). Nasal secretion, however, was significantly increased at the higher relative humidity.

At levels between 0.1 and 1 ppm formaldehyde the frequency of complaints increased slowly but steadily with increasing exposure (4). At levels of exposure greater than 1 ppm the dose-response curve became much steeper; there were about twice as many complaints at 3 ppm as at 1 ppm and generally over four times as many complaints at 5 ppm as at 1 ppm. At all concentrations the frequency of complaints rapidly decreased with time. After 20 min the plateau of all complaints for the group exposed to 5 ppm was approximately that of the 3 ppm group. After 40 min of exposure the frequency of symptoms was about the same in the 3 ppm exposed group as in all groups with lesser exposure; the greatest decrease in complaints occurred in the first 20 min of exposure. All the groups exposed to $\leqslant 1$ ppm formaldehyde had a similar frequency of complaints after 10-15 min of exposure. The adaptation to formaldehyde seen in the exposure chamber was also observed in anatomy students exposed in the laboratory.

Andersen (5) exposed 16 subjects for 5 h to 0.25, 0.42, 0.83, and 1.6 ppm formaldehyde at 50 percent relative humidity. Subjective discomfort, nasal resistance, VC, FEV_1, $FEF_{25-75\%}$, odor threshold, and nasal mucociliary flow were measured. The odor threshold for ethyl valerate was significantly increased at 1.7 ppm but not at any of the lower concentrations of formaldehyde. Symptoms (mainly eye irritation and dryness in nose and throat) were the most sensitive indicators. Discomfort was dose related but was slight even at the highest exposure. In all groups, maximum discomfort was reached in 2-4 h; at the 1.7 ppm exposure some subjective adaptation occurred, as the mean discomfort score declined in the last half of the exposure period. The proportion of subjects registering complaints was 19, 31, 94, and 94 percent at 0.3, 0.4, 0.8, and 1.7, respectively. All symptoms had completely disappeared by the next morning. There were never any lower airway symptoms.

Animal Studies

Kane and Alarie (6) evaluated sensory irritation of formaldehyde in mice by measuring changes in respiratory rate. Many species (including humans and mouse) show a characteristic reflex decrease in respiratory rate when exposed to upper respiratory irritants. The respiratory rate decreased 50 percent at 3.1 ppm $(1.5 - 6.7 = 95$ percent C.I.) and 30 percent at 1 ppm. After 10 min the response plateaued and then began to decrease. When mice were exposed for 3 h over a 4-day period, the same pattern was observed, except at 3.1 ppm on the fourth day the decreased response was delayed. At both 1 ppm and 3.1 ppm exposure concentrations the maximum decrease in respiratory rate was greater on each succeeding day, suggesting the development of hypersensitivity on continued exposure.

Alarie has proposed a quantitative relationship between percent decrease in respiratory rate in mice and eye, nose, and throat irritation in humans (Table 1). The dose-response relations in mice for sensory irritation are quite close to those observed in the controlled exposure of human subjects discussed above.

Population Studies

What has been the experience outside of the laboratory, where exposure is less well defined and is usually to a mixture of pollutants rather than to formaldehyde alone (Table 2)?

Bourne and Seferian (7) reported burning and stinging of the eyes, nose and throat irritation, and headache among customers in dress shops who were exposed to 0.13-0.45 ppm formaldehyde.

Blejer et al. (8) reported on workers in a garment factory where the fabric was impregnated with a resin composed mostly of urea-formaldehyde and glyoxal (α-dialdehyde). Breathing zone concentrations of formaldehyde ranged from 0.9-2.7 ppm in different areas of the plant. There was a pungent, readily noticeable odor where larger quantities of work material accumulated, and it was in these areas that the greatest discomfort (lacrimation, nose and throat irritation) occurred. The irritation was greatest when beginning work and again after returning from lunch; 15-20 min after initial exposure the irritation became tolerable. The irritation was greatest in the winter when it was foggy, humid, or cold, but this could be due to higher concentrations as ventilation was poorer in colder weather.

Kerfoot and Mooney (9) reported eye and nose burns, sneezing, coughing, and headaches among seven embalmers in six funeral homes. Three of the seven workers suffered from asthma or sinus problems, and two had dermatitis.

Table 1 Quantitative Relationship

ppm HCOH	Decrease in respiratory rate (RD_{50})	Irritation in humans (extrapolated)
3.1	50% decrease in respiratory rate in mice	Severe and intolerable irritation, rapidly incapacitating
0.31	0.1 RD_{50} in mice	Slight but tolerable sensory irritation in humans
0.03	0.01 RD_{50} in mice	Minimal or no sensory irritation in humans

Formaldehyde vapor concentrations ranged from 0.09-5.26 ppm; the range of average concentrations was 0.25-1.39 ppm. Formaldehyde concentrations were approximately doubled when the ventilation system (fan) was off.

Plunkett and Barbela (10) also found a high prevalence of eye irritation (81 percent), nose and throat irritation (75 percent), and cough (33 percent) among 57 of 80 (71 percent) embalmers responding to a self-administered mail questionnaire. Other symptoms relating specifically to formaldehyde exposure included skin irritation (37 percent), chest tightness (23 percent), shortness of breath (10 percent), and wheezing (12 percent). The prevalence of acute bronchitis and chronic bronchitis was 16 percent and 30 percent, respectively. About 80 percent of those with bronchitis were smokers or exsmokers, compared to 52 percent of the asymptomatic workers.

A NIOSH health hazard evaluation of an embalming laboratory at a college of mortuary science revealed formaldehyde levels above 1 ppm (one personal sample for 4 h was 3.9 ppm) when the ventilation system was inoperative (not an unusual situation) (11). When the ventilation system was operating all samples were less than 1 ppm. Phenol air concentrations were below the limit of detection. One 30-year-old instructor had worked about 27 h/week for 10 years and had an early disability retirement because of asthmatic bronchitis that may have been work related. Three of the four current instructors gave positive histories of allergies and all reported symptoms of burning eyes and nose, dryness of the mouth and throat, cough, headache, and lacrimation during days of higher exposure. The four instructors had worked 3 weeks to 13 years for 8, 12, 21, and 50 h/week.

Embalming fluid contains 1 percent or more formaldehyde, but it has other ingredients, including paraformaldehyde or polyoxymethylene, paradichlorobenzene, tissue moisturizers, smooth muscle relaxants, bleaching and auxiliary antiseptic agents (phenol), dyes, buffers and salts, wetting agents, water conditioners and/or anticoagulants, perfumes and odor suppressors, and vehicles (methanol, ethanol, glycerine).

The results of these three studies of embalmers are only suggestive, as sample sizes are biased and small, and there were no control groups; environmental data were lacking on one study. A large, well-designed study of this occupational group should be done.

Wayne and co-workers (12) examined the eyes and ascertained symptoms in 83 workers employed in a particleboard and plywood manufacturing plant. The frequency of complaints was generally high: bodily fatigue (52 percent), headache, sore throat, cough, stuffy or runny nose (41 percent), eye discomfort (62 percent), respiratory or eye discomfort attributed to unsatisfactory air (38 percent), eye fatigue (21 percent), odor (41 percent). There was no apparent difference in complaints among the four exposure groups (high = 0.56-1.41 ppm; third quartile = 0.3-0.55 ppm; second quartile = 0.15-0.29 ppm; first quartile = 0.03-0.14 ppm). Eye discomfort was the most frequent complaint;

Table 2 Summary of Epidemiological Studies of Irritation

Reference	Estimated exposure	Type of exposure	Symptoms	Comment
7	0.1–0.5 ppm	Customers in dress shops	Burning and stinging of eyes, nose, and throat irritation, headache	
8	0.9–2.7 ppm	Workers in garment factory (urea-formaldehyde resin)	Lacrimation, nose and throat irritation	Dose-related adaptation in 15–20 min
9	0.1–5.3 ppm	7 embalmers in 6 funeral homes	Eye and nose burns, sneezing, coughing, headache	Dose-related; no control group
10		57 of 80 embalmers	Eye irritation (81%), nose and throat irritation (75%), cough (37%), skin irritation (37%), chest tightness (23%), shortness of breath (10%), wheezing (12%)	Self-administered questionnaire; no control group; no environmental measures
11	> 1 ppm No phenol detected	4 embalmer instructors	Burning of eyes and nose, dryness of mouth and throat, cough, headache, lacrimation	NIOSH Health Hazard Evaluation; no symptoms when < 1 ppm; early disability retirement with asthmatic bronchitis of instructor with 5 years' exposure
12	0.03–1.41 ppm	83 employees of particle-board and plywood manufacturing plant	Bodily fatigue (52%), headache, sore throat, cough, stuffy or runny nose (41%), eye discomfort (62%), respiratory or eye discomfort attributed to unsatisfactory air (38%), eye fatigue (21%), odor (41%)	Not dose-related

	Concentration	Population	Findings	Comments
14	0.1-3 ppm (most < 1)	168 mobile home residents complaining of formaldehyde (102 > 18 years old)	Eye, nose, throat irritation (78%), cough and wheeze (38%), respiratory problems (30%), diarrhea (28%), headache (52%), nausea and vomiting (20%), skin rash (0)	Mean HCHO concentration highest in homes of those with eye, nose, throat irritation; concentration related to age of mobile home; no control group; investigated complaints only
15	N.D.-2.5 ppm (91% < 1 ppm)	Residents of 68 mobile with complaints (n = 239)	Eye irritation (34%), nose irritation (5%), respiratory tract (24%), headache (21%), nausea (5%), drowsiness (11%)	Not dose-related; no control group; investigated complaints only
16		395 UF-foam insulated homes; 400 control homes	*Prevalence* (exposed vs control): Asthma (1.1% vs 0.9%), wheezing (2.0% vs 0.9%) $p = 0.02$, burning skin (0.8% vs 0.2%), headache (1.9% vs 3.2%) $p = 0.04$, burning eyes (2.1% vs 2.6%), cough (2.7% vs 3%), symptomatic persons (15% vs 17%), symptomatic households (33% vs 35%) *Incidence* (exposed vs control): Wheezing (0.6% vs 0.1%) $p = 0.01$, burning skin (0.7% vs 0.1%) $p = 0.04$, symptomatic persons (9.9% vs 9.9%), symptomatic households (23% vs 24%) *Incidence* (preinsulation versus postinsulation) (#/1000): asthma (1.2 vs 0.9), wheezing (1.2 vs 2.3), burning eyes (2.4 vs 5.5), runny nose (3.1 vs 3.2), cough (1.2 vs 12.4), headache (3.1 vs 3.7), symptomatic persons (15.3 vs 45.5), symptomatic households (21 vs 70)	No obvious differences between exposed and controls; prevalence much lower than previous studies No obvious differences between exposed and controls; incidence of symptoms twice as high in exposed households reporting *any* odor compared to exposed households reporting *no* odor Tendency for incidence of symptoms in preinsulation group to be higher than controls; dose-response in that postinsulation incidence higher than preinsulation

Table 2 Summary of Epidemiological Studies of Irritation (*Continued*)

Reference	Estimated exposure	Type of exposure	Symptoms	Comment
17	<1 ppm with periodic increases	278 wood processing workers (extensive use of U-F resins); 200 age-matched controls	*Prevalence* (exposed vs controls): chronic rhinitis (14–36% vs 3%), chronic pharyngitis (1.5–13% vs 1%), chronic laryngitis (0–3.5% vs 1.5%), chronic sinusitis (0–4.2% vs 1.5%), nasal cilia depressed and olfactory threshold elevated in exposed compared to controls	Majority of workers complained of throat irritation, decreased olfaction, and dryness in nose and pharynx; rhinitis and pharyngitis were more prevalent in workers < 5 years than ⩾ 5 years tenure
18	0.44 ppm 0.57 ppm 0.13 ppm	1594 students in three schools with U-F chip plates compared with 497 students in a school with no HCHO-emitting chip plates	Children in exposed schools had significantly higher prevalence of functional disturbances (headache, disorder of concentrating ability, dizziness, nausea), respiratory tract irritation (nose, pharynx, dry cough), conjunctiva irritation, and other complaints (somnipathy, abdominal pain, skin disease)	In one school where emission sources were removed, complaints were reduced 71%

25 percent complained before beginning work, 44 percent during work, and 54 percent after the work shift.

The Consumer Products Safety Commission has received hundreds of complaints about products containing formaldehyde. About 75 percent were related to urea-formaldehyde foam, 20 percent to mobile homes, and about 6 percent to particleboard, carpets, drapes, clothes, and newsprint. Health complaints included headache, dizziness, nausea, conjunctivitis, coryza, coughing, dyspnea, and rashes.

Several studies have involved the residents of homes insulated with formaldehyde resins (13–15). Two studies involved the investigation of complaints from owners of mobile homes. Garry et al. (14) found that mean formaldehyde concentrations were higher in the homes of persons with index symptoms (eye, nose, and throat irritation) and that persons with a history of asthma tended to report symptoms at a lower formaldehyde exposure. Most exposure levels were less than 1 ppm. Breysse (15) did not show any apparent association with exposure, which ranged from < 0.22 ppm to 2.54 ppm with 90 percent of the sample < 1 ppm. There was no comparison group in these studies.

In a New Jersey study reported by Thun (16) there was a control group, although no formaldehyde measures were taken. Telephone interviews were conducted with an adult member of the household of 395 homes insulated with urea-formaldehyde (UF) foam and 400 control homes in the same neighborhood that had not been insulated with UF foam. Information collected included demographic information, data on the construction and insulation of the home, formaldehyde odor, and symptoms and other medical data. The foam and control samples were well matched on age, sex, family size, and smoking. Analysis was in three parts:

1 Prevalence data (all symptoms regardless of when they began). Of 13 symptoms, the foam population had a higher rate for only 3 (asthma, wheezing or difficulty breathing, and burning skin). The difference was statistically significant only for wheezing (2 percent versus 0.9 percent). The controls reported a statistically significant increase of headache (3.2 versus 1.9 percent).

2 Incidence data (symptoms that began within the study year). The foam population had a higher incidence of asthma, wheezing, chest pain, burning skin and eyes, and sore throat. The differences were significant only for wheezing (0.6 versus 0.1 percent) and burning skin (0.7 versus 0.1 percent). The controls had a higher incidence (not significant) of runny nose, rash, and dizziness. The incidence was the same for cough, headache, insomnia, and vomiting.

3 Incidence of symptoms before and after UF foam installation. The postinsulation incidence of symptoms was higher than the preinsulation incidence, and the latter group had slightly lower incidence than the control population. None of the differences was striking because of the infrequent acquisition of symptoms in all populations. Among the foam households reporting *any* formal-

dehyde odor the incidence of symptoms was twice that of foam households reporting *no* odor. The incidence was dose related, and the persistent odor group reported seeking more medical attention.

The authors point out several limitations and potential biases of this important study. An unknown number of insulated homes were not a part of the sample. Health effects were based only on telephone interviews, and the interviewee had to recall symptoms over a year period. A potential bias between exposed and control populations was possible on the reporting of symptoms. With these limitations in mind, the following tentative conclusions were made:

1   There did not appear to be an overall excess of morbidity in the UF foam-exposed population.

2   There appeared to be an increase in symptoms following installation of insulation.

3   The greatest evidence of a health effect occurred in that part of the population (28 percent) that reported odor, and particularly the group with persistent odor (9 percent). Unfortunately it was not possible to determine if this group represented a population with poor health, chronic overreporters, or hypersusceptibles.

Yefremov (17) examined the upper respiratory tract of 278 wood processing workers and 200 age-matched workers not involved in formaldehyde production. There was extensive use of urea-formaldehyde resins, and although formaldehyde levels were periodically elevated, the author states that workroom formaldehyde did not deviate significantly from maximum permissible concentrations (= 1 ppm). The prevalence of rhinitis was elevated in all exposure groups (14-36 percent) compared to controls (3 percent). Pharyngitis was elevated only in the higher exposure groups. There were no detectable differences in the prevalence of laryngitis and sinusitis. Although older workers had a higher proportion of rhinitis and pharyngitis, the prevalence was higher among those working less than 5 years compared to those working 5 years or more (16 percent versus 2.6-5.2 percent for rhinitis), suggesting selection. However, the reverse was seen for physiological measures, as exposed workers with longer tenure had a higher prevalence of reduced nasal clearance, increased olfactory thresholds, and anosmia.

Burdach and Wechselberg (18) examined 1594 students in three schools built with formaldehyde-emitting chip plate and 497 students in a school without the chip plates. Formaldehyde averaged 0.44, 0.57, and 0.13 ppm in the three schools. Functional disturbance (headaches, concentrating ability, dizziness, nausea), respiratory tract and conjunctival irritation, somnipathy, abdominal pain, and skin disease were more frequent among the formaldehyde-exposed children. Removal of emission sources in one of the schools resulted in a 71 percent reduction in complaints.

Summary

In summary, there is no doubt that formaldehyde is an upper respiratory tract and mucous membrane irritant. Exposures over 10 ppm are self-limiting because of the severe discomfort (profuse lacrimation, difficulty in taking normal inspiration, burning of nose, throat, and trachea, coughing) (19). Levels of 4-5 ppm can be tolerated for 10-30 min by some people. OSHA has a permissible exposure level of 3 ppm; ACGIH recommends a limit of 2 ppm based on irritation of eyes, respiratory tract, and skin (20); and NIOSH has recommended a ceiling level of 1 ppm for any 30-min sampling period based on odor and irritation (21). Most environmental and industrial exposures are probably around 3 ppm or less. It is also in this range that adaptation or acclimatization appears to occur. Irritant dose-response relations in this range of exposures are summarized in Table 3.

PHYSIOLOGICAL EFFECTS (TABLE 4)

There are at least three mechanisms by which formaldehyde might cause asthmatic types of responses:

1 Reflex bronchoconstriction. Airflow obstruction can occur in healthy people on inhalation of irritant substances (e.g., SO_2, NH_3). If a high proportion of the exposed population show airflow obstruction and symptoms, it is reasonable to infer that the inhalant is a primary irritant and all exposed are at risk. Whether this occurs in humans on exposure to formaldehyde vapor is not documented, and there is some evidence it does not. Airflow obstruction does occur in humans on exposure to particulates containing formaldehyde and in guinea pigs exposed to formaldehyde vapor.

Table 3   Irritant Dose-Response Relationships for Airborne Formaldehyde Exposure

Sign or symptom	Concentration (ppm)	Rank of frequency at 0.5 ppm
Odor	0.05 (0.025-1)	1
Eye irritation	0.01-1.2	2
Nose irritation	0.05-1.2	3
Throat irritation	0.05-2.1	3
Lacrimation	0.05	4
Slight discomfort	0.03-1.3	
Increased rate of eye blinking	1.7	
Increased odor threshold	1.7	
Reduced nasal mucus flow	0.25	
Increased nasal secretion	0.2	5
Dryness of the throat	0.2	5

Table 4 Summary of Acute and Chronic Effects of Formaldehyde Exposure

Estimated formaldehyde exposure	Length of exposure	Number of subjects	Type exposure	Findings	Reference
7.5 ppm	2 min	2–7	Mouthpiece	No change in airway resistance, FEV_1%	4
0.3, 0.4, 0.8, 1.6 ppm	5 h	16	Exposure chamber	No change in FC, FEV_1, FEF_{25-75}%. Decrease in nasal mucus flow	5
Usually less than 10 ppm	≥5 yr	25	Making urea-formaldehyde and phenol-formaldehyde resins	No chest X-ray findings related to formaldehyde exposure	23
<4.1 ppm	7 yr	18	Storage and processing of crease-resistant cloth	Normal VC; 1 of 18 with hyper-irritable airways by acetylcholine provocation test (unrelated to formaldehyde)	24
0.5 ppm 0.4 mg/m³ dust	≤18 mo	52	Hexamethylene tetramine-resorcinol adhesive	No apparent reduction in spirometry or increase closing volumes related to exposure	25
0.5 ppm 0.4 mg/m³ dust	20 min to 3–5 h	19	Hexamethylene tetramine-resorcinol adhesive	Reduction in FEF_{50} (−6%) and FEF_{75} (−17%) associated with particulate but not formaldehyde exposure; no change in other spiro-metric parameters or closing volume	25
0.4–13.6 ppm HCHO; no phenol detected	>5 yr 1–5 yr <1 yr	15 10 40 15	Felting acrylic-wool filters, impregnating with liquid phenol-formaldehyde resin, and then drying	Excess cough (35%) and cough plus phlegm (25%) compared to pre-viously and never exposed groups; no apparent reduction in FVC or FEV_1; smoking adjusted FEV_1% and FEF_{50}/FVC reduced in groups with more than 1 yr exposure	26

Not measuring during acute study	Mon AM–Mon PM Mon AM–Fri AM Mon AM–Fri PM	12	Felting acrylic-wool filters, impregnating with liquid phenol-formaldehyde resin, and then drying	Tendency toward increased acute symptoms (lower respiratory tract symptoms, cough, phlegm, wheezing/chest tightness, shortness of breath); slight decrease in pulmonary function over workday and workweek	26
1.86 mg/m³ phenolic resin dust	Workshift	36	Phenol-formaldehyde resin powder added to processed cotton by garnet-line workers	Reduction in FEV_1 (-122 ml) and FVC (-125 ml)	27
1.9–3.1 mg/m³ respirable phenolic	<5 yr (ever smoked) ≥5 yr (ever smoked)	31 (ever smoked) 20 (ever smoked)	Workers who had ever worked at job with exposure to phenolic resin dust; include workers with exposure to both phenolic resin and processed cotton dust (3.1 mg/m³)	Smokers: Workers with ≥5 yr increased prevalence of FEV_1 <80% (10% vs 3%, N.S.), FEV_1 % <80% (15% vs 0%, N.S.); years worked significant association with FEV_1 % (-0.5/yr), exposure variable N.S. for all other parameters	27
≥0.3 ppm	1 h		Guinea pigs	Increased airway resistance; decreased compliance	28
3.7 ppm	24/hr/day/90 days	15	Rats, guinea pigs, rabbits, dogs, monkeys	Interstitial inflammation of lungs with uncertain relation to formaldehyde exposure	29
0.1 ppm HCHO 4 mg/m³ NaCl	1 h	13	Guinea pigs	Increased airway resistance (no change in resistance when HCHO was less than 0.1 ppm)	28
0.02, 0.2, 2 ppm	6 h	23	Young pigs	No change in compliance, arterial blood gases, radiology of thorax	30

2   Immunologic. The characteristics of an allergic mechanism are that the response can be evoked in sensitized individuals with very small amounts of formaldehyde. Symptoms usually develop some time after the initial exposure rather than on initial contact. Usually only a proportion of the exposed will be affected. In the documented cases there is typically a delayed response, although there may be a brief immediate reaction as well (dual response), and the late reaction may be prolonged.

3   Pharmacologic. The response depends on the release of a mediator. Formaldehyde may directly stimulate histamine release from mast cells, the histamine causing the symptoms. Nicholls (22) showed that dimethylhydantoin-formaldehyde resin releases histamine from guinea pig, monkey, and human lung. Whether this mechanism is operative in humans is not known.

Controlled Studies

Rader (4) attempted to determine the effects on respiratory mechanics of exposure to 7.5 ppm formaldehyde, a level at which coughing did not occur. Airway resistance showed no apparent decrease in two subjects after inhalation for 2 min of 7.5 ppm formaldehyde. $FEV_1\%$ (FEV_1/FVC) also showed no apparent reduction in three subjects. Similar results were obtained in five subjects who were less skilled. The variance was quite large in the unskilled subjects and the power was quite small with this small number of subjects. Airway resistances and $FEV_1\%$ measure primarily obstruction of airflow in the trachea, bronchi, and larger airways. Both are relatively insensitive measures of acute airflow obstruction. Apparently the formaldehyde was also breathed only through the mouth, thereby effectively bypassing nasal receptors.

Anderson (5) observed no apparent effect on FVC, FEV_1 and $FEF_{25-75\%}$ after 5 h exposure of 16 subjects to formaldehyde concentrations up to 1.6 ppm. However, the sample size was small, so that there was an 80 percent chance of correctly rejecting the hypothesis of no differences between control and exposed at the 10 percent α level when the true differences were about 600 ml for FEV_1, 700 ml/s for $FEF_{25-75\%}$, 650 ml for VC, and 9 mm H_2O. Nasal mucus flow was reduced in the anterior third of the nose at all exposure concentrations (except 0.83 ppm). The reduction was greater at 4 h (as much as a 50 percent decrease) than at 2 h. There was no apparent reduction of mucus flow in the posterior two-thirds of the nose.

Population Studies

Harris (23) examined 25 men making urea-formaldehyde resins and powders and some phenol-formaldehyde resin. Each man had worked at least 5 years and had presumably been exposed daily to formaldehyde. General plant levels were usually less than 10 ppm. Four of the workers had had mild cases of dermatitis.

Four had mild dyspnea on exertion (one had hypertension and one asthma). Chest X-rays showed no pattern of findings that could be related to the formaldehyde exposure.

Kratochvil (24) examined 18 workers exposed to formaldehyde in the storage and processing of crease-resistant cloth. Formaldehyde was less than 4.1 ppm. Average exposure was 7 years. The group was small, and the author found normal values for vital capacity. One person had hyperirritable airways as measured by the acetylcholine bronchoprovocation test, but this was considered to be a result of acute catarrh of the bronchi following influenza.

Gamble et al. (25) administered spirometry, single breath O_2, and a respiratory questionnaire to 157 rubber workers; 52 exposed to a hexamethylene tetramine-resorcinol (H-R) adhesive system, 50 workers in similar jobs where the H-R system was not used, and 55 other workers not exposed to the H-R adhesive system. There was no demonstrable difference in cough, phlegm, or baseline lung function (FEV_1, FVC, flows, closing volume) among the three exposure groups. Acute changes in lung function and symptoms associated with the job were also determined in 19 H-R exposed men, 16 of the H-R exposed group on a day when they were not working with the H-R system, and 19 nonexposed controls. Area samples for formaldehyde were similar in the three groups (0.5 ppm, 0.2 ppm, and 0.4 ppm, respectively), but only the H-R exposed group had significant reductions in flow rates ($FEF_{50} = -6\%$, $FEF_{75} = -17\%$) and they were associated with respirable particulate exposure. FEV_1 and FVC were not reduced over the shift. H-R exposed workers had significantly increased rates of acute symptoms (itch, rash, breathing better away from work, cough, chest tightness, burning eyes, runny nose, and burning sensation in heart region).

Schoenberg and Mitchell (26) administered respiratory questionnaires and spirometry to three groups of employees in a filter manufacturing plant: 40 production line workers exposed to 0.4–0.8 ppm formaldehyde (measured with indicator tube samples) on one occasion and 8.8–13.6 ppm (three breathing zone samples) on another occasion when fans were not operating during part of the sampling period; 8 men who had previously worked on the production line; and 15 men who had never been production line workers (3 in this group were maintenance workers and were exposed two to three times per month to high formaldehyde concentrations). The exposed group of production line workers had more chronic cough and chronic cough and phlegm than the other groups. After adjustment for smoking, the 15 men in the exposed group who had worked more than 5 years had reduced FEF_{50} and $FEV_1/FVC\%$; those working 1–5 years had intermediate values. There was no demonstrable difference in FEV_1 and FVC between the groups.

Acute symptoms were also determined in these three groups (nonexposed, previously exposed, currently exposed) (26). The prevalence was high in all groups, over half reporting eye, nose, and throat irritation, lower respiratory tract symptoms, and cough. Most subjects reported getting the symptoms shortly after a short high exposure, with the symptoms going away in 5-15 min. Seven subjects reported a delayed onset (several hours to many hours after the end of the shift) of symptoms (particularly wheezing) after initial exposure. The delayed symptoms usually lasted for several hours to 2 days. Wheezing, chest tightness, and shortness of breath were higher (38 and 45 percent) in the exposed group compared to previously exposed (12 and 25 percent) and never exposed (7 and 7 percent). None of the differences was significant, however, because of the small sample sizes.

Despite the high prevalence of symptoms, acute changes in lung function were small (26). Pulmonary function over the shift and the week showed a tendency to decrease in a subset of exposed population ($n = 12$) and to increase in a subset of controls ($n = 11$). The only significant change was in the exposed group with a -1.6 percent difference in FEV_1 between Monday morning and Friday evening. Changes in lung function over the shift were also small. The authors suggest exposure was low on the day of the test, although environmental measures were not taken on the days when spirometry was performed. The authors conclude that formaldehyde exposure may result in an increase in symptoms and airflow obstruction.

Sparks and Peters (27) studied respiratory morbidity in 73 men and women exposed to phenolic resin, to processed cotton dust, or to both. Respirable dust concentrations of phenolic resin dust were high: 1.86 ± 1.5 mg/cubic meter for seven samples. For the study of acute changes in lung function, 36 garnet line workers exposed to resin dust wore respirators. Mean reductions in FEV_1 and FVC were -122 ml and -125 ml respectively, both significantly different from zero. Baseline pulmonary function of workers ever exposed to resin dust were compared to published normal regression equation. The proportion with percent predicted FEV_1, FVC, and $FEV_1\%$ less than 80 percent was higher in the group working 5 years or more than in the group working less than 5 years, but the differences were significant for FEV_1 only. When pulmonary function was analyzed by multiple regression among those who had ever smoked, the years worked variable was small but significant for percent predicted $FEV_1\%$ (-0.47 percent/year) and the years smoked variable (-0.55 percent/year smoked) for FEV_1. No other variable showed a statistically significant association nor were there significant associations among nonsmokers.

The authors suggest that exposure to phenolic resins has both an acute and a chronic effect on pulmonary function. While phenolic resins are polymers of phenol and formaldehyde, the actual formaldehyde content of the dust and thus exposure to formaldehyde is not known and was probably mixed to some extent with cotton dust. The respirable dust exposure was associated with substantial acute reductions in FEV_1 and FVC, but the association of exposure with reductions in baseline FEV_1 and FVC was not convincing.

Animal Studies

Short-term (1 h) air exposure of guinea pigs to concentrations of formaldehyde as low as 0.31 ppm resulted in increased airway resistance and decreased compliance (28). Longer term exposure (24 h/day/90 days) at 3.7 ppm formaldehyde resulted in interstitial inflammation of the lungs but no clinical signs of illness or toxicity in 15 rats, 15 guinea pigs, 3 rabbits, 2 dogs, and 3 monkeys (29). One rat died. The authors were not certain the changes were caused by the formaldehyde, but suggested further studies were necessary because of the limited nature of their experiment and the one animal death.

Frey et al. (30) exposed 23 young pigs for 6 h to 0.02, 0.2, and 2 ppm formaldehyde. They observed slight alterations in lung structure of the 2 ppm exposed group but no other difference in lung function (compliance), arterial blood gases, or radiology of the thorax.

Summary

The controlled studies of humans exposed to formaldehyde vapor show no acute effect on respiratory mechanics. However, sample sizes were small, and such an effect cannot be ruled out. The results in the epidemiological and animal studies are not the same. In the epidemiologic studies, there were other potential irritants, the formaldehyde was a particulate, and particulate carriers were available. The epidemiologic studies are the only ones examining chronic effects of exposure. The results are not consistent, and the effects are small where there were positive results. More studies are required to confirm the acute and chronic effects of formaldehyde on lung function.

At least two mechanisms for airflow obstruction seem relevant. Irritation of upper airway receptors could result in airflow obstruction, although this has not been documented for formaldehyde vapor in humans. Whether a reflex bronchoconstriction could eventually cause chronic airflow obstruction is not known. The second mechanism could be the direct action of particulate formaldehyde (e.g., resin) on formaldehyde adsorbed onto a particulate carrier. Amdur (28), for example, has shown a synergistic response of formaldehyde with a NaCl carrier in guinea pigs. The lack of an acute reduction in lung function in humans when exposed to formaldehyde as a vapor and reductions where the formaldehyde exposure has been to a particulate suggest that the second mechanism may be operative in humans.

SENSITIZATION (TABLE 5)

Formaldehyde is a known sensitizer and numerous cases of allergic contact dermatitis have been described (21). Pulmonary sensitization, however, is less often described, and Fassett (19) in 1963 knew of no "instances of authentic

Table 5   Summary of Cases of Formalin Asthma

Reference	Exposure	Response	Confirmed by formaldehyde challenge
32	Fur workers, shoe and rubber industry, disinfection	Bronchial asthma in 1 subject; dyspnea at end of shift and at night; positive reaction to formaldehyde patch test and inhalation challenge; no eosinophilia, P-K reaction; normal spirometry	Delayed reaction
33	Formalin from a hemodialysis unit; frequent spills	Persistent dry cough/episodic attacks of wheezing that developed into productive cough, persistent wheezing, increasing breathlessness, and rhinitis; inflammatory changes on X-ray	Delayed reaction
34	Formalin from a hemodialysis unit; frequent spills	Attacks of wheezing, early dry cough that became productive before onset of wheezing; attacks provoked by heavy Formalin exposure, becoming more frequent and severe; 1–2 wk required for full recovery	Delayed reaction
37	Catalytic spray paint releasing formaldehyde particles	Runny nose for 3 yr after spray painting; severe asthmatic attack requiring hospitalization following spray painting under crowded conditions; nocturnal symptoms following attack; stopped spray painting and became asymptomatic; moderate bronchial hyperreactivity to histamine	Peak flow reduced in 3 h following exposure to paint
35	Formalin fixative (40% HCHO) 15 hr in high concentrations week before; 2 h exposure before onset, HCHO noted on breath	Irritation of nose and eyes; progressive dyspnea and chest tightness over 15-h period following exposure; on day 2: dyspnea at rest, X-ray with increased interstitial markings with early edema, reduced FEV_1, FVC, and MMEF with some improvement 3 wk later; asymptomatic 5 wk later	Not challenged; diagnosed as either acute chemical pneumonitis or a hypersensitivity reaction in an atopic individual
36	Formalin fixative	Exposure to Formalin vapor at work led to attacks of cough and severe chest wheezing; several times severe bronchial asthma; asymptomatic when away from work	Not challenged; diagnosed as late-onset bronchial asthma of nonallergenic type

pulmonary sensitization." Formaldehyde is often referred to as causing occupational asthma (31), and a few authenticated cases have been described.

Popa et al. (32) described several cases of occupational Formalin asthma in fur workers, tanners, and in workers in shoe and rubber industry and in disinfection. All had delayed reactions, and some were sensitive to paraphenylene diamine or bichromate or both as well as Formalin. Shortness of breath began 4-6 h after initial exposure, remained for 1-2 h, subsided, and then returned in the night. However, inhalation provocation tests were positive in 1 of 6 cases where Formalin was thought to be the only sensitizing agent, and none (0/20) where Formalin was one of the chemicals and paraphenylene diamine, bichromate, or both were the others. The authors suggested that Formalin was acting as an irritant in those subjects in whom there was no signs of sensitization by either skin test or inhalation. Positive skin tests in 2 of the subjects were thought to represent a latent allergy. However, it is not clear why they would show a delayed response if formaldehyde were acting as an irritant. The one case with bronchial asthma and a positive inhalation test to nonirritating concentrations of Formalin showed a delayed hypersensitivity reaction.

Hendrick and Lane (33, 34) described recurrent episodes of productive cough and wheeze in 8 of 28 hospital workers exposed to formaldehyde from a hemodialysis unit. Although no environmental measurements were taken, exposures were apparently high and continual, with some undiluted Formalin spills producing even higher short-term exposures. Four staff members and a sister of a patient, all of whom had a history of asthmatic symptoms since becoming regularly exposed to formaldehyde, were administered inhalation provocation tests. The tests were carried out when subjects were asymptomatic, and spirometry was close to predicted normal values. In two of the nurses, inhalation provocation tests with formaldehyde reproduced the symptoms seen at work. Wheezing began 2-3 h after exposure and reached a maximum 6-22 h later. Four hours after initial exposure, peak flow decreased. Productive cough was a prominent symptom. The duration of the response was related to dose and in some instances had lasted for weeks. A threshold for the asthmatic response seems likely, as at a lower concentration no response was elicited from one of the nurses. The other three individuals did not show a specific response to Formalin. The authors suggested that continued exposure to Formalin vapor may increase susceptibility to bronchitis attacks or make the airways hyperreactive.

Two other reported cases of Formalin-induced respiratory disease in laboratory workers are suggestive of hypersensitization (35, 36). In at least one of the workers, exposures must have been quite high as formaldehyde could be detected on his breath (35).

Alanko et al. (37) reported 5 occupational asthma cases (8 percent of the total number of cases in Finland during 1975) wherein the etiologic agent was considered to be formaldehyde. In Finland, formaldehyde has recently become an important allergen, with most of the work-related asthma cases caused by formaldehyde from glues used in the veneer and plywood industry. They describe a case in which the occupational exposure was spray painting, with the release of formaldehyde particles. Spray painting under crowded conditions resulted in a serious asthmatic attack requiring hospitalization. The patient is symptom-free except when around paint. Peak flow decreased 42 percent when provoked with paint (with a continual reduction 3–12 h after exposure) and Formalin (night attack 16 h after exposure and a 47 percent reduction in peak flow). The individual had no asthma on exertion, no eosinophilia, negative scratch test, and moderate bronchial hyperactivity to histamine. He was a smoker with no history of infantile eczema or allergic rhinitis and no family history of atopy.

Occupational asthma has been estimated to comprise 2 percent of all cases of asthma, although the prevalence varies widely by country and occupation (38). Although the prevalence of formaldehyde-induced asthma is unknown, in sensitized subjects it does cause severe and prolonged attacks of asthma, even at low exposure concentrations. The mechanisms of the respiratory reactions and sensitivity are not known. The irritant properties of formaldehyde are the more common responses. There are reports suggesting an association between formaldehyde and airflow obstruction. However, further documentation is needed, particularly for the vapor, to determine consequences of both acute and chronic exposure. While Formalin asthma is not commonly reported in the literature, the incidence of new cases may be on the increase because of the increasing use of formaldehyde. The number of confirmed cases of formaldehyde asthma is more than reported in the literature. In Finland, for example, there are many confirmed (but unreported) cases. In one case, the subject was exposed to particleboard in the home. When provoked with less than 1 ppm formaldehyde there was a delayed response of asthmatic symptoms and airflow obstruction (Nordman, H, personal communication). When hypersensitivity does occur, exposure can result in episodes that are quite severe and long-lasting. Studies to determine the incidence and prevalence of Formalin asthma are needed.

NIOSH has recommended a 30-min exposure ceiling of 1 ppm (21). Below 1 ppm the irritant effects of formaldehyde are minimal, and after several minutes adaptation occurs so they are no longer noticeable. In a sensitized individual there should be no exposure to formaldehyde.

CANCER

The report of nasal tumors in rats exposed to formaldehyde has within the past year raised a concern about the carcinogenicity of formaldehyde. Despite its widespread use and the large numbers of people exposed to formaldehyde, there are few data on its potential carcinogenicity. The most important study is that of Swenberg et al. (see Chap.12).

If formaldehyde is a human carcinogen, the site and degree of toxicity may be modified from that observed in rats. Rats are obligatory nose breathers, and so all the formaldehyde is deposited in the nose. In humans, the deposition sites would depend on the proportion of nose and mouth breathing. Another unanswered question relates to the ability of the human respiratory tract to metabolize and detoxify the formaldehyde.

SUMMARY

The most common effect of formaldehyde on the respiratory system is irritation. Eye, nose, and throat irritation can occur before odor detection. Irritation is slight at concentrations less than 1 ppm, and adaptation at these levels occurs in minutes. The frequency and severity of irritation increases rapidly at levels above 1 ppm, and adaptation is neither as rapid nor as complete the greater the exposure. Although formaldehyde has been shown to produce airflow obstruction in guinea pigs, there are no controlled human exposures wherein airflow obstruction is produced by formaldehyde vapor. Several epidemiological studies of workers exposed to formaldehyde or formaldehyde resin dust suggest both acute and chronic airflow obstruction, but more research is needed. The actual formaldehyde exposure is not known.

The most convincing evidence for a toxic effect of formaldehyde on the lung are the cases of Formalin asthma. Bronchial provocation in sensitized individuals produces delayed asthmatic responses (e.g., airflow obstruction, wheezing) that may last for several weeks. The prevalence of pulmonary sensitization in exposed populations is not known. There are no published studies in human populations to adequately assess the carcinogenic potential of formaldehyde.

REFERENCES

1 Sim, VM, and RE Pattle. Effects of possible smog irritants on human subjects. *JAMA* 165:1908–1913 (1957).

2 Schuck, EA, ER Stephens, and JT Middleton. Eye irritation responses at low concentrations of irritants. *Arch. Environ. Health* 13:570–575 (1966).

3 Weber-Tschopp, A, T Fischer, and E Grandjean. Reizwirkungen des Formaldehyds (HCHO) auf den Menschen. *Int. Arch. Occup. Environ. Health* 39:207–218 (1977).

4 Rader, J. Irritative effects of formaldehyde in laboratory halls. J.D. Dissertation, University of Wurzburg (1974).

5 Andersen, I. Formaldehyde in the indoor environment—health implications and the setting of standards. In *Indoor Climate*, Proceedings of the First International Indoor Climate Symposium, edited by P. O. Fanger and O. Valbjorn. Copenhagen, Denmark, August 30–September 1, 1978.

6 Kane, LE, and Y Alarie. Sensory irritation to formaldehyde and acrolein during single and repeated exposure in mice. *Am. Ind. Hyg. Assoc. J.* 38: 305–322 (1977).

7 Bourne, HG, and S Seferian. Formaldehyde in wrinkle-proof apparel produces . . . tears for milady. *Ind. Med. Surg.* 28:232–233 (1959).

8 Blejer, HP, BH Miller, and JT Ganotes. Occupational health report of formaldehyde concentrations and effects on workers at the Bayly Manufacturing Company, Visalia. State of California, Department of Public Health, Study Report Number S-1806, April 5, 1966.

9 Kerfoot, EJ, and TF Mooney. Formaldehyde and paraformaldehyde study in funeral homes. *Am. Ind. Hyg. Assoc. J.* 36:533–537 (1975).

10 Plunkett, MD, and T Barbela. Are embalmers at risk? *Am. Ind. Hyg. Assoc. J.* 38:61–62 (1977)

11 Johnson, PL, B Froneburg, and D O'Brien. *Health Hazard Evaluation.* Determination Report Number HE 79-146-670. Cincinnati College of Mortuary Science Embalming Laboratory, Cincinnati, Ohio (1979).

12 Wayne, LG, RJ Bryan, and K Ziedman. *Irritant Effects of Industrial Chemicals: Formaldehyde.* US DHEW, PHS, CDC, NIOSH, Publication Number 77-117 (1977).

13 Sardinas, AV, RS Most, MA Giulietti, and P Honchar. Health effects associated with urea-formaldehyde foam insulation in Connecticut. *J. Environ. Health* 41:270–272 (1979).

14 Garry, VF, L Oatman, R Pleus, and D Gray. Formaldehyde in the home. *Minn. Med.*, pp. 107–111 (Feb. 1980).

15 Breysse, P. Formaldehyde in mobile and conventional homes. *Environ. Health Safety News* 26:1–6 (1977).

16 Thun, MJ, MF Lakat, and R Altman. New Jersey urea-formaldehyde foam insulated study, mimeographed report. New Jersey State Health Department.

17 Yefremov, GG. The upper respiratory tract in formaldehyde production workers. *ZH UShn. Nos. Gorl. Bolezn.* 30:11–15 (1970).

18 Burdach, S, and K Wechselberg. Damages to health in schools. Complaints caused by the use of formaldehyde-emitting materials in school buildings. *Fortschr. Med.* 98:379–384 (1980).

19 Fasset, DW. Aldehydes and acetals. In *Industrial Hygiene and Toxicology,* Vol. 2, 2d ed, edited by FA Patty, pp. 1959–1989. New York: Interscience (1963).

20 American Conference of Governmental Industrial Hygienists. *Documentation of the Threshold Limit Values for Substances in Workroom Air,* pp. 118–119. Cincinnati: ACGIH (1971).

21 *Criteria for a Recommended Standard . . . Occupational Exposure to Formaldehyde.* US DHEW, PHS, CDC, NIOSH (1976).

22 Nicholls, PJ. Release of histamine from lung tissue in vitro by dimethylhydantoin-formaldehyde resin and polyvinyl pyrrolidane. *Br. J. Ind. Med.* 33:127–129 (1975).

23 Harris, DK. Health problems in the manufacture and use of plastics. *Br. J. Ind. Med.* 10:255–268 (1953).

24 Kratochvil, I. Effect of formaldehyde on the health of workers employed in the production of crease-resistant clothing. *Pracouni lekarstvi* 23:374–375 (1971).

25 Gamble, JF, AJ McMichael, T Williams, and M Battigelli. Respiratory function and symptoms: An environmental/epidemiologic study of rubber workers exposed to a phenol-formaldehyde type resin. *Am. Ind. Hyg. Assoc. J.* 37:499–512 (1976).

26 Schoenberg, JB, and CA Mitchell. Airway disease caused by phenolic (phenol-formaldehyde) resin exposure. *Arch. Environ. Health* 30:574–577 (1975).

27 Sparks, MJ, and JM Peters. Respiratory morbidity in workers exposed to dust containing phenolic resin. *Int. Arch. Occup. Environ. Health* 45:221–229 (1980).

28 Amdur, MO. The response of guinea pigs to inhalation of formaldehyde and formic acid alone and with a sodium chloride aerosol. *Int. J. Air Pollut.* 3: 201–220 (1960).

29 Coon, RA, RA Jones, LJ Jenkins, and J Siegel. Animal inhalation studies on ammonia, ethylene glycol, formaldehyde, dimethylamine and ethanol. *Toxicol. Appl. Pharmacol.* 16:646–655 (1950).

30 Frey, G, KH Back, H Meister, HV Hang, J Kilion, and FW Ahnefeld. Effects of ventilation with defined formaldehyde concentrations on lung function and lung structure. *Anaesthesist* 28:281–278 (1979).

31 Brooks, SM. Bronchial asthma of occupational origin. *Scand. J. Work Environ. Health* 3:53–72 (1977).

32 Popa, V, D Teculescu, P Stanescu, and N Gabrilescu. Bronchial asthma and asthmatic bronchitis determined by simple chemicals. *Dis. Chest.* 56:395–404 (1969).

33 Hendrick, DJ, and DJ Lane. Formalin asthma in hospital staff. *Br. Med. J.* 1:607–608 (1975).

34 Hendrick, DJ, and DJ Lane. Occupational formalin asthma. *Br. J. Ind. Med.* 34:11–18 (1977).

35 Porter, AH. Acute respiratory distress following formalin inhalation. *Lancet* 2:603–604 (1975).

36 Sakula, A. Formalin asthma in hospital laboratory staff. *Lancet* 2:816 (1975).

37 Alanko, K, H Keskinen, and L Saarinen. Occupational asthma. *Duodecim* 93:306–318 (1977).

38 Davies, RJ, and J Pepys. Occupational asthma. In *Asthma*, edited by TJH Clark, and S Godfrey. Philadelphia: Saunders (1977).

Chapter 17

Symptom Survey of Residents of Homes Insulated with Urea–Formaldehyde Foam: Methodologic Issues

Michael Thun

Ronald Airman

New Jersey became involved in the urea-formaldehyde (UF) foam controversy because of a steady and troubling flow of complaints and questions from consumers about the product. Like other governmental agencies, we found ourselves pressured to provide information and to adopt a regulatory stance toward a product about which we know very little. We were aware of reports of skin, eye, and respiratory tract irritation among some households exposed to the foam. Early in 1978, we knew of three families who were unable to use all or part of their home because of persistent irritating odor.

Because we had little confidence in our surveillance system, however, we felt that there were three basic public health questions to be answered in order to better define the problem and to develop appropriate regulation. 1) How common are problems (health, structural, and odor) in a population of insulated homes? 2) How severe are these problems? 3) Why are some homes affected and not others; or to state the question differently, could we epidemiologically characterize the distinguishing features of the high-risk subgroup?

The focus of this chapter is not the results of the study that we conducted in New Jersey, since these have been submitted for publication elsewhere.

Instead we will address some general methodologic issues that pertain to field research on formaldehyde. Our experience in New Jersey has made us intimately aware of some of the problems and controversies surrounding research into formaldehyde toxicity.

STUDY DESIGN

The study design that one selects to answer public health questions is almost always chosen based on two considerations: a) What is the question to be answered? and b) What resources are available to answer it? The resources that we had available in New Jersey to answer the three questions mentioned above were limited. The most fundamental limitation was that we had only certain mechanisms available to identify a study group. The preferred mechanism, one that was not available to us, would have been some system that identified homes as they were insulated. Such a system would have permitted us to identify UF foam-exposed homes before they were insulated and to measure formaldehyde concentrations before and after installation of the foam. A prospective study design could define the amount of formaldehyde that UF foam contributes to the domestic environment, the natural history of off-gassing, and, most importantly, the occurrence of effects in relation to given levels of formaldehyde exposure.

Instead, the resource available to us was an opportunity to look retrospectively at the experience of UF foam-insulated homes. We received lists from four UF foam manufacturers of all the homes in New Jersey that had been insulated with their product. Although these lists were not epidemiologically ideal, since they did not represent all manufacturers, and their completeness could not be verified, they were an important start.

STUDY LIMITATIONS

However, with the adoption of a retrospective approach, several intrinsic limitations of the study were unavoidable. The most important was our inability to conduct timely air monitoring. Because most of the homes had been insulated a number of months before we could identify the study population from the lists, no objective measurements of formaldehyde exposure immediately following insulation could be obtained. Instead, we were reliant on indirect, subjective indicators of exposure. The most readily available was the description of odor, as reported by the household occupants. Reports of odor duration following installation of the foam varied widely between families (Fig. 1). A subgroup of concern was the 8.4 percent of households reporting odor of $\geqslant 7$ days of duration. However, subjective odor perception is an imperfect measure of exposure, subject to interobserver variability, diminished sensitivity to odor with prolonged exposure, and subject bias.

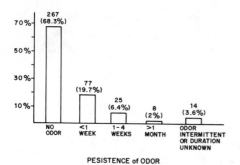

PESISTENCE of ODOR

Figure 1 Percentage of households by answer to odor duration question. Based on 391 households who answered the odor question. (*New Jersey 1979 UF Foam Study.*)

A second constraint, not directly related to the retrospective design, was our inability to assess health effects by objective means. Instead, we were reliant on symptom reporting in response to a telephone-administered questionnaire.

Symptom surveys have a number of problems, some of which are intrinsic to the technique, others of which have special importance when evaluating toxins with irritative effects such as formaldehyde. For certain health effects, symptom surveys are completely insensitive. Obviously, the tool is of no value at all to evaluate subclinical effects, chronic effects with long latency, or health problems that are either not perceived or not reported by persons involved. In view of the serious health questions about the chronic effects of formaldehyde, such deficiencies are substantial.

Second, reporting of symptoms is susceptible to memory bias. The more distant a person is from a health event, the less likely she/he is to remember or report it. The strong correlation between symptom reporting and recency has been described previously (1) and was evident in our study in the reporting of new symptoms by neighborhood control households (Fig. 2). In comparing the

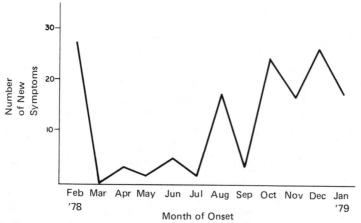

Figure 2 Number of new symptoms per month reported by controls. (*New Jersey 1979 UF Foam Study.*)

exposed and nonexposed populations, one must compare equivalent, and preferably recent, time periods.

The most important limitation of symptom surveys, from a practical standpoint, is subject bias. A subject may underreport or overreport symptoms based on his or her preconceptions about the exposure under study. This bias is of greater concern in situations in which the subject is necessarily aware of his or her exposure status. A single- or double-blinded study design is unattainable in evaluating a product such as UF foam insulation. The problem of subject bias is compounded in a retrospective study, in which the bias may influence assessment of both exposure and effect. To obtain definitive answers, it is necessary to use measurements of exposure, effect, or both, which are independent of subject bias.

A final limitation of the symptom survey in assessing domestic formaldehyde exposure is the nonspecificity of many of the acute effects of formaldehyde. None of the symptoms of eye, skin, or respiratory tract irritation is specific to formaldehyde exposure. All have a high background prevalence in the community. If one uses the presence or absence of symptoms to separate affected "cases" from unaffected persons, one may blur the distinguishing features of those persons whose problems are truly related to formaldehyde exposure. High background noise may obscure both a real effect of the foam and the characteristics of those homes and persons who develop adverse reactions to the foam.

CONCLUSION

Thus, the solution is to pursue research designs in which assessment of either exposure or effect is independent of subject bias and in which the effect is highly specific for formaldehyde. Collection of better exposure data requires cooperation between industry and government, making available study populations in which domestic formaldehyde exposure can be measured. In addition, objective measurements of the biological effect of formaldehyde should be evaluated for field use. One such possibility is testing for formaldehyde-specific humoral precipitating antibody as an indicator of hypersensitivity (2). Pulmonary function testing, measuring airway changes in exposed populations, is another potentially useful tool. Tests of pulmonary function have proved logistically difficult to conduct in non-occupational field settings. However, if a number of homes in close geographic proximity could be identified, such testing might prove feasible. Patch testing has been used extensively to detect hypersensitivity among dermally exposed occupational populations. It remains unclear whether patch testing is a useful experimental tool for evaluating individuals who have received predominately respiratory exposure.

SUMMARY

We have discussed several methodological issues pertaining to studies of formaldehyde-related morbidity in the non-occupational setting. We have attempted to place the symptom survey conducted in New Jersey in its appropriate context, as a relatively crude instrument, useful in assessing the magnitude and immediacy of a public health problem. Our purpose in focusing on the limitations of the study was not to undercut its value but rather to emphasize that it is only a beginning. It should not be taken out of context, nor should it be interpreted as evidence of the nonpathogenicity of domestic formaldehyde exposure. Objective measurements of exposure and more specific indicators of biological effect are needed to better define the effects of such exposure.

REFERENCES

1 U.S. Department of Health, Education and Welfare, National Center for Health Statistics. *A Summary of Studies of Interviewing Methodology*. Rockville, MD: NCHS Series 2 No. 69 (1977).
2 Oziganova, VA, IS Ivanova, and LA Deuva. Bronchial asthma in radio equipment operators. *Sov. Med. (Moscow)* 8:139–141 (1977).

Chapter 18

Formaldehyde in the Home Environment: Prenatal and Infant Exposures

Mary Ann Woodbury

Carl Zenz

Do home air contaminants cause acute infant illness? Are infants more susceptible than adults to home air contaminants? These are the questions that have been raised since the first reports of such infant illnesses were made in 1978 to the Wisconsin Division of Health.

This is a case history report of infant illnesses in homes wherein a noxious air contaminant was suspected to be present. Summary statistics will be used to describe these cases. In examining these data, it should be recognized that these infants constitute a case series rather than a representative sample of infants with such exposures. It should also be recognized that the cases were not compared to a control group nor were formaldehyde vapor measurements taken in subsequent environments where illness did not occur.

METHODOLOGY

Requests for assistance were obtained from physicians, the public, or other agencies between January 1, 1978 and November 1, 1979. In each case, a home

We would like to give our thanks to Kay A. Dally, Industrial Hygienist and Lawrence P. Hanrahan, Research Analyst for their assistance on this project.

air contaminant was suspected as a cause of illness. The cases described are limited to those we investigated. If the symptoms of upper respiratory and eye irritation were reported in one or more residents for any period of time in a home with newer urea-formaldehyde (UF) resin products, an industrial hygienist made a site visit. During the site visit, detailed information was collected on the home. Samples for formaldehyde vapor were collected in the breathing range in each home by conventional air sampling techniques utilizing the National Institute for Occupational Safety and Health (NIOSH) chromotropic acid procedure (1).

If there was a question of an outside formaldehyde vapor source, outside air samples were taken. (Such samples were taken on only two occasions and both outdoor samples were below detectable limits.) Each adult (age 18 and older) in the home completed a detailed health questionnaire. A parent completed a questionnaire for each child over 12 months of age. A staff interviewer administered a separate infant questionnaire for each child under 12 months of age. In all cases of acute infant illness, one of us (C.Z.) discussed the case with the attending physician.

GENERAL FINDINGS

Findings of what were presumed to be formaldehyde-related illnesses were investigated in 92 homes. Sixty-five of these homes were mobile homes, 14 were conventional homes with UF foam insulation, and 13 were conventional homes with UF wood products. All but one of the infant cases were in infants residing in mobile homes. The number of structures* by the age of the building materials in the home is given in Table 1. There is a statistically significant relationship between the age of the building materials and the level of the formaldehyde found in the home ($p < 0.005$): the older the building materials the lower the level of formaldehyde. The association between cigarette habits by

---

*Several buildings and recreational vehicles in addition to homes were investigated and are included in Table 1.

Table 1   All Formaldehyde Complaint Units (of Any Type) by Age of Building Materials

Formaldehyde concentration (ppm)	Building materials < 15.8 months of age	Building materials > 15.8 months of age
<0.10	6 (30%)	14 (70%)
0.11–0.34	6 (21%)	23 (79%)
0.35–0.80	20 (64%)	11 (36%)
>0.81	16 (80%)	4 (20%)

Chi^2 for trend = 21.77; $p < 0.005$.

Table 2 Smoking Status Inside Dwellings by Concentration of Formaldehyde

Formaldehyde concentration (ppm)	Smokers absent	Smokers present
<0.10	14 (78%)	4 (22%)
0.11–0.34	12 (52%)	11 (48%)
0.35–0.80	18 (62%)	11 (38%)
>0.81	8 (53%)	7 (47%)

Chi^2 for trend = 1.50; $p > 0.10$.

household was also investigated. No association between households with smokers and the level of formaldehyde in the air was shown ($p > 0.10$) (Table 2).

There were 260 people in the 92 investigated homes. The age distribution of the complaintants was not unusual. The male to female ratio was approximately equal, although in the infants, males slightly outnumbered females 1.5 to 1.0. Table 3 indicates the symptoms for the people for whom health histories were available. History of chronic disease or asthma before this exposure, 21 percent and 6 percent, respectively, was not unusual. Eye irritation was the only

Table 3 Symptoms Reported for All Occupants of Complaint Homes by Order of Frequency

Symptom or condition	No	Yes	% Yes
Eye irritation	81	165	67*
Runny nose	100	146	60
Dry or sore throat	106	140	57
Burning eyes	100	146	59
Headaches	116	130	53
Cough	122	124	50
Difficulty sleeping	153	93	38
Diarrhea	164	82	33
Nausea	170	76	31
Phlegm	180	64	26
Weakness	191	53	22
Vomiting	195	50	20
Dizziness	197	48	20
Wheezing	198	47	19
Chest tightness	203	42	17
Breathlessness	205	41	17
Rash	205	40	16
Chest pain	215	29	12
Bronchitis	225	20	8
Pneumonia	232	13	5

*Statistically significant $p < 0.0001$.

symptom that could be statistically correlated to formaldehyde level with a simple chi^2 test ($p < 0.0001$). A trend-type association was also found to be statistically significant for eye irritation ($p < 0.005$): the higher the level of formaldehyde in the home the more complaints of eye irritation. The chi^2 for trend, a more sensitive test, might show a significant association for other reported symptoms (those tests have not been done).

Twenty infants under 12 months of age with postnatal exposure to formaldehyde are included in this series. Their symptoms by order of frequency are given in Table 4. The most common symptoms were restlessness, red eyes, diarrhea, and vomiting. Rhinorrhea and cough were also frequently reported. Although problems with breathing were only reported on a frequent or constant basis in 5 cases, it must be recognized that Table 4 reflects the frequency not the severity of illness. For instance, apnea cannot be a constant problem, but it is a life-threatening problem when it occurs. In all of these cases, the parents or physicians or both reported cessation of these symptoms when the child was out of the home for a period of time. Physician confirmation of this pattern was obtained in the majority of cases. In all cases of acute illness, diagnostic workup failed to yield any possible etiology for the illness. In some of the infants, the vomiting was projectile in nature and necessitated evaluation for pyloric stenosis.

These infants, after being symptom free out of the home, again developed symptoms when returned to the home. In one case, the family physician had the family return the infant to the home three times; each time diarrhea reappeared, after 8 h on the second return and after 2 h on the third return. It must be emphasized, there was no vomiting or diarrhea in this infant when he was out of the home for periods of 2-30 days. A similar pattern was seen in the other cases.

Table 4   Symptoms[*] Reported for Twenty Infants[†]

Symptom	Never or occasionally	Frequently or constantly
Restlessness	11	9 (45%)
Red eyes	11	8 (42%)
Diarrhea	12	8 (40%)
Vomiting	12	8 (40%)
Rhinorrhea	12	7 (37%)
Cough	13	7 (35%)
Difficulty sleeping	14	6 (30%)
Constant crying	14	6 (30%)
Excessive sleeping	14	5 (26%)
Difficulty breathing	15	5 (25%)
Sluggish/lethargic	15	5 (25%)

[*]When symptoms were present in four or more infants.
[†]Twelve or less months of age.

Out of 20 infants with postnatal exposure, 9 required hospitalization 16 times. One infant, hospitalized for a urinary infection, was not classified as a hospitalization in this analysis as it is unlikely to be related to formaldehyde. Four of the hospitalized infants were admitted for gastrointestinal problems (diarrhea, vomiting), and five were hospitalized for respiratory problems (rales, tachypnea, pneumonia, respiratory distress syndrome). Two of the infants hospitalized for respiratory problems also experienced frequent gastrointestinal problems. One of the four hospitalized for gastrointestinal problems was also hospitalized for pneumonia on one occasion.

Eleven of the infants underwent 7 or more months of gestation while the mother lived in the formaldehyde-contaminated home. The mean birth weight for the 11 infants was 7 lb, 8 oz (standard deviation 1 lb, 11 oz). Gestation lengths were all within normal limits. Of the 11 infants, one was anencephalic and had minor birth defects and a low birth weight (3 lb, 11 oz) with a normal length gestation. Three of the eleven mothers worked full time outside of the home during pregnancy and thus were away from exposure for a minimum of 45 h per week. Two mothers worked part time during pregnancy. Neither mother worked outside of the home for the two birth defect cases. The two infants with birth defects were from two of the three highest levels of exposure, 0.54 ppm for the anencephalic child and 2.76 ppm for the other child.

The median formaldehyde level for all the infant cases was 0.33 ppm. The range of formaldehyde vapor levels was 0.05–2.76 ppm. Table 5 gives the median, geometric mean, arithmetic mean, and the standard deviation (SD) for formaldehyde levels for various infant groups. In each group, 2 or 3 cases with very high exposures have skewed the data. The geometric mean is a better indicator of central tendency because of this skewed distribution.

In comparing the hospitalized infants versus nonhospitalized infants, the median values are quite different in the expected direction (1.9–1.0), although it is somewhat less if the geometric mean is utilized (1.3–1.0).

The small numbers make interpretation difficult in comparing the infants by reason for hospitalization. While the median and the arithmetic mean indicated that infants hospitalized for gastrointestinal illness had higher exposures, the geometric mean indicated the reverse. The small numbers and the small differences make it impossible to comment on the direction of the association.

The median level for the 11 maternal exposures was 0.27 ppm, with a geometric mean of 0.36 ppm. As previously indicated, two of the highest exposure levels were in the homes with the birth defect cases. When those 2 cases are removed, the nine normal infants had a mean maternal exposure of 0.29 ppm. The removal of those 2 cases has reduced the variation in the data substantially (SD 0.75–0.12) and left the median and geometric and arithmetic means quite similar. These home measurements were taken for some cases prenatally and for others postnatally. For the 2 birth defect cases, the 0.54 ppm level was found in sampling prior to birth in the winter and the 2.76 ppm level

Table 5   Formaldehyde Vapor Levels[*] Median, Means and Standard Deviation for Various Infant Groups

Group	Number	Median	Geometric mean	Arithmetic mean	Standard deviation
All exposed infants[†]	20	0.33	0.44	0.68	0.69
Hospitalized infants	9	0.51	0.46	0.70	0.62
Respiratory[‡]	5	0.51	0.52	0.68	0.63
Gastrointestinal[‡]	4	0.60	0.39	0.73	0.71
Nonhospitalized infants[§]	9	0.27	0.35	0.41	0.26
Prenatally exposed	11	0.27	0.36	0.54	0.75
Prenatally exposed without birth defect cases	9	0.26	0.28	0.29	0.12

[*]All formaldehyde measurements are given in parts per million.

[†]Not including those infants who were prenatally exposed but never in the contaminated home.

[‡]Primary reason for hospitalization. (One child classified GI was hospitalized twice for GI and once for pneumonia.)

[§]Excluding birth defect case.

was found in sampling 2 months after birth on a hot, humid August day. The 0.54 level was in the home of the anencephalic child, while the 2.76 reading was in the home of the child with low birth weight and minor defects. If formaldehyde was responsible for these birth defects, this seeming discrepancy between exposure and effect may be the result of the prevailing weather conditions at the time of the sampling, i.e., average exposures during the pregnancies of these two women may not be reflected by these samples.

None of the infant symptoms showed a statistically significant association with formaldehyde vapor level in the home. The hospitalization rates were not statistically significant when compared by formaldehyde level.

DISCUSSION

Food, water, and microbial agents seem unlikely factors in these cases because of the nature of the home dependence phenomena. Other than air and water, the possible sources of exposure move with the family, including the smoking habits. Water-borne illness could be considered but city water supplies and no use of water for some of the infants would seem to make this unlikely. At the same time, the pattern of infant illnesses and onset of illness appears to be related to changes in heat, humidity, and use of windows. As the urea-formaldehyde resin is well known to break down (hydrolyze) with changes (increases) in temperature and humidity, the apparent weather/temperature-related pattern of these cases tends to lend further evidence to the role of airborne formaldehyde in these illnesses. For another agent to have been responsible, it would have to have similar changes with heat and humidity. A microbial agent could fit this pattern, but we believe it is unlikely to fit both the weather and the home dependence phenomena. Diagnostic tests have failed to yield any pathologic organism and infants have had elevated temperatures only in cases of pneumonia.

Other organic and inorganic air contaminants also need to be considered. Nitrogen and sulfur oxides can be present at low levels in homes with gas stoves (2, 3). However, the usual exposures of the nitrogen and sulfur oxides are substantially below industrial levels (by a factor of 100 or more compared to 4–10 for formaldehyde). One formaldehyde exposure, 2.76 ppm, was almost at the maximum allowable industrial level (4), and three homes exceeded the standard recommended by the National Institute for Occupational Safety and Health for the workplace (5). Despite the extremely low nitrogen dioxide levels in these homes, two studies of pulmonary function in children from homes with gas stoves do suggest that these levels of contaminants may be sufficient to cause physiologic effects in children (6, 7). Although nitrogen and sulfur oxides may play a role, the typical gastrointestinal problems have occurred in homes with electric appliances. At the same time, the weather-related pattern of these cases would not tend to support a causal role for gas stove-related contaminants. Spot checks for carbon monoxide were done in two of the homes in our infant series; no detectable carbon monoxide was found. Other organic pollutants are certainly present and may also play a role in these cases. In general, these homes are quite tight, with low air exchange levels. Because of the multiple pollutants present in homes and the low levels of ventilation, the possibility of synergistic effects must be considered.

Ordinarily, when questions of the effects of chemicals are raised, animal studies and occupational studies would be given heavy consideration in evaluating potential human hazards. In this situation, two problems are present relative to the infant and such comparisons. Occupational studies concern healthy adults exposed 8 h a day, 5 days a week. Because exposure is affected by body weight, respiratory rate, volume of inspired air, length of exposure, body surface area, ability to metabolize the substance, rate of metabolism, and developmental status (i.e. age), a continuous exposure of an infant cannot be considered comparable to that of an adult worker. Animal studies have been done only on weanling animals and cannot be considered comparable to human infants under 12 months of age. In regard to the teratology studies, not only have such studies not simulated a continuous exposure, but inadequate attention has been addressed to species differences in the metabolism of formaldehyde and to the effects of folate status (8–10). The monkey or folate-deficient rats would appear to be the only appropriate animal models for adequate studies of the teratogenic potential of formaldehyde.

We have not addressed or discussed the special vulnerability of the premature or the disabled infant who may be brought into a polluted home. This concern and the noted respiratory effects also raise the question of the role of home air contaminants in the sudden infant death syndrome. Apnea has been noted in some animals as a reflex reaction (superior laryngeal nerve) to chemical irritants (11). One of the infants in this series required electronic monitoring for apnea.

CONCLUSION

To return to the original question, home air contaminants do cause acute infant illnesses. While formaldehyde alone may not be conclusively associated with such infant illnesses, the apparent temporal association between temperature, humidity, ventilation, age of home, and illness all support the causal role of formaldehyde. Acute illness did not occur in any of these infants' parents. It appears that infants are more sensitive than adults to airborne contaminants. However, there may have been differences in the length of continuous exposures between parents and infants that may explain the apparent differences in health effects. Because of the case history nature of these cases, the lack of adequate controls, and the need for measurement of other contaminants, controlled research is needed to substantiate our suggestion that formaldehyde plays a causal role in these cases.

REFERENCES

1 National Institute for Occupational Safety and Health (NIOSH). *Manual of Analytical Methods.* DHEW Pub. #77-157, 125-1 (1977).

2 Wade, WA, WA Cote, and JE Yocom. A study of indoor air quality. *Air Pollut. Cont.* 25:933–939 (1975).

3 Spengler, JD, BG Ferris, and DW Dockery. Sulfur dioxide and nitrogen dioxide levels inside and outside homes and implications on health effects research. *Environ. Sci. Technol.* 13:1276–1280 (1979).

4 Occupational Safety and Health Administration. General industry standards. *Fed. Register*, 29 CFR 1910-1000 (1975).

5 National Institute for Occupational Safety and Health. *Criteria for a Recommended Standard . . . Occupational Exposure to Formaldehyde.* DHEW (NIOSH) Pub. #77-126 (1976).

6 Melia, RJW, C deFlorey, and S Chinn. The relation between respiratory illness in primary school children and the use of gas cooking. *Epidemiology* 8:333–353 (1979).

7 Speizer, FE, F Ferris, YMM Bishop, and J Spengler. Pulmonary function in children associated with NO_2 exposure. *Am. Rev. Resp. Dis.* 121:3–10 (1980).

8 Clay, KL, RC Murphy, and WD Watkins. Experimental methanol toxicity in the primate: Analysis of metabolic acidosis. *Toxicol. Appl. Pharmacol.* 34:49–61 (1975).

9   McMartin, KE, G Martin-Amat, PE Noker, and TR Tephly. Lack of a role for formaldehyde in methanol poisoning in the monkey. *Biochem. Pharmacol.* 28:645–649 (1979).

10 Makar, AB, and TR Tephly. Methanol poisoning in the folate-deficient rat. *Nature* 261:715–716 (1976).

11 Oliver, TK, and WM Burnett, eds. *Neonatal Respiratory Adaptation.* Proceedings of the Interdisciplinary Conference on Neonatal Respiratory Adaptation, p. 134. USDHES (NICH & HD) Pub. #1432 (1963).

Chapter 19

The Effects of Occupational Exposure on the Respiratory Health of West Virginia Morticians

Richard J. Levine

R. Daniel Dal Corso

Patricia B. Blunden

Mario C. Battigelli

Embalmers disinfect, preserve, and restore bodies in preparation for funerals. During the course of work they are exposed principally to the toxic effects of formaldehyde and its polymers, which constitute the primary active ingredients of embalming fluids, jellies, and powders.

Formaldehyde is known to cause skin and upper respiratory irritation,

This work was funded by the Chemical Industry Institute of Toxicology, a not-for-profit research organization operated in the public interest and supported by the chemical industry.

We are grateful to Mr. V. Mancinelli of the West Virginia Board of Embalmers and Funeral Directors for his enthusiastic support of the study and to him and Mr. O. Fansler of the Department of Pathology, West Virginia University Medical Center, for making many necessary arrangements. We thank Dr. M. Symons for statistical consultation, Mr. S. Spain for assistance with the pulmonary function measurements, and the following persons for their participation in the field survey: C. Chiles, T. Chin, G. Decad, D. Dodd, N. Fitch, D. Guest, R. Krieger, M. Mabry, and J. Sun. We would like to express special appreciation to Dr. R. Krieger for help in securing a high participation rate in the survey. Finally we acknowledge with gratitude the use of facilities of the Department of Pathology, West Virginia University Medical Center, and the loan of spirometers from the University of North Carolina School of Medicine and the National Institute for Occupational Safety and Health.

allergic eczematous contact dermatitis, and possibly asthma; however, the potential for formaldehyde to induce chronic changes in pulmonary function is unclear. Small airway flow rates among rubber workers exposed to a mixture of respiratory irritants including formaldehyde were observed to diminish over the course of a workshift; however, no chronic effects could be detected (1). On the other hand, length of employment in an industrial department that produced acrylic-wool filters impregnated with a phenol-formaldehyde resin was associated with reduced Monday preshift FEV_1/FVC and FEF_{50}/FVC,* but no acute functional losses were reported (2). It has been suggested that embalmers and persons who work in renal dialysis units, where artificial kidneys are sterilized with Formalin, have an increased frequency of acute and chronic bronchitis (3, 4). To help resolve these matters, the authors examined the respiratory status of 105 white male morticians attending a postgraduate school of instruction in Morgantown, West Virginia.

BACKGROUND

Two important categories of embalming practice can be distinguished in relation to exposure: the embalming of intact bodies and of autopsied bodies. Six West Virginia funeral homes in the vicinity of Morgantown were surveyed, and personal air samples were taken over the length of time actually required for embalming. Time-weighted-average formaldehyde concentrations were reported to be 0.3 ppm for intact bodies and 0.9 for autopsied bodies. Peak half-hour concentrations were 0.4 ppm and 2.1 ppm, respectively (5). While exposures encountered in embalming autopsied bodies are more intense, autopsied bodies usually comprise only a minority of total bodies embalmed. The actual proportion is heavily influenced by location and clientele.

Embalming the Normal Case

Wearing an apron and rubber gloves, the embalmer places the subject onto a metal embalming table, which is inclined slightly from the horizontal. After the subject has been bathed and shaved and the mouth and eyes have been closed, massage cream is applied to the face, neck, ears, and hands to prevent desiccation. The body is arranged in a natural position, but the head, feet, and elbows are elevated to facilitate distribution of embalming fluids and drainage of blood. Embalming fluids are diluted with water in an open tank and infused into a suitable artery. Provided there is no obstruction to flow, fluids injected into any artery of reasonable size readily circulate throughout the vascular system and diffuse into the tissues. In order to prevent distention of the vasculature and tissue discoloration from extravasated blood, venous blood with admixed

---

*FEV_1 = one-second forced expiratory volume; FVC = forced vital capacity; FEF_{50} = forced expiratory flow rate at 50% FVC.

embalming fluid is allowed to flow onto the embalming table and drain. A jet of water may be placed at the upper end of the table to wash the surface.

Thoracic and abdominal viscera cannot be embalmed by arterial infusion because of low infusion pressure and the rapid postmortem deterioration of organ capillaries. Embalming of the viscera is accomplished by thrusting a trocar through the lower abdominal wall and diaphragm to puncture the internal organs, to suction blood and tissue fluid, and to instill a concentrated preservative called "cavity fluid." To minimize leakage from the abdomen following withdrawal of the trocar, a plastic button is screwed into the subcutaneous fat at the insertion site. A few hours later, after the cavity fluid has had sufficient time to set the organs, the button is removed and the trocar reintroduced to suction remaining fluids and gases.

Embalming the Autopsy Case

At autopsy the thoracic and abdominal vasculature are disrupted by incision of the aorta and removal of the viscera. Each limb and each side of the head, therefore, must be infused and drained separately. Although severed vessels are clamped with vascular forceps, blood and embalming fluid drain into the body cavities and require suctioning. A preservative is injected into the cavity walls via a trocar or large-bore needle. Interior cavity surfaces are coated with embalming jelly or powder. Viscera are removed from the body, washed, bathed in preservative fluid, and returned to the body after embalming has been completed. If the brain has been autopsied, sealing compound is applied to the incised intracranial vasculature to prevent leakage, and an embalming powder is spread over the interior surface of the skull.

Embalming Agents

Besides formaldehyde and its derivatives, a variety of chemicals may be employed in embalming fluids. These include tissue moisturizers, such as glycerol, sorbitol, lanolin, and glycols; smooth muscle relaxants, such as squill, nitroglycerin, magnesium chloride, and tetrapotassium pyrophosphate; bleaching and auxiliary antiseptic agents such as phenol; dyes, such as saffranine, methyl red, Congo red, and eosin; buffers and salts, such as sodium bicarbonate, sodium carbonate, potassium nitrate, magnesium sulfate, boric acid, borax, sodium chloride, disodium phosphate, and oxalates; wetting agents, such as sodium lauryl sulfate and sodium hexametaphosphate; water conditioners and anticoagulants, such as ethylenediaminetetraacetic acid and sodium citrate; perfumes and odor suppressants, such as oil of cloves, sassafras, oil of wintergreen, benzaldehyde, oil of organe flower, lavender, and rosemary; and vehicles, such as methanol, ethanol, and glycerine (6). Arterial injection fluids usually are diluted to a formaldehyde concentration of 1 or 2 percent, but higher concentrations are employed in cavity fluids and for treating problem cases.

The principal active ingredients of embalming powders are the formaldehyde polymers paraformaldehyde and polyoxymethylene. In addition to these,
embalming powders may also contain plaster of Paris, alum, salt, phenol, wood
powder, clays, and paradichlorobenzene (6).

Restorative Art

The embalmer may wish to restore exposed portions of the body in order to
achieve a desired cosmetic effect. The usual case requires only sufficient dye
in the arterial embalming fluid or the application of routine cosmetics to im-
part lifelike hues. The hair must be brushed and combed and may be kept in
place with hair tonic or spray. Special situations may call for corrective surgery;
plaster of Paris and wax may be used to fashion body parts; bleaches and solvents may be applied to remove stains; astringents may be injected to reduce
distention; and sunken areas may be restored to a normal contour by the injection of "tissue builder" cream (7).

METHODS

In order to remain licensed, West Virginia morticians are required to attend one
postgraduate school of instruction every 3 years. Schools of instruction are held
twice a year in different parts of the state. On April 18, 1979, 105 of 112 (94
percent) licensed white male morticians attending a school of instruction in
Morgantown completed self-administered respiratory disease questionnaires and
a detailed occupational history; 99 of these (88 percent) underwent pulmonary
function testing.

Questionnaires

The ATS-DLD-78 respiratory disease questionnaires recommended by the
American Thoracic Society and the National Heart-Lung Institute Division of
Lung Diseases (8) were followed by questions on the upper respiratory tract and
occupational history. Subsequent telephone interviews were conducted to elicit
responses to questions that had not been answered on the day of testing.

A subject was defined to have chronic bronchitis if positive responses were
obtained to independent queries concerning cough and bringing up phlegm on
most days for 3 or more consecutive months during the year.

Smoking histories permitted classification of subjects into groups who had
never smoked (nonsmokers), those who had quit smoking (exsmokers), and
current smokers (smokers). Data were converted to an index of pack-years of
exposure by multiplying the average number of cigarettes smoked per day times
the number of years smoked, divided by 20. Pipe and cigar smokers were considered to be nonsmokers providing they inhaled not at all or slightly and had
never smoked cigarettes. Smokers of pipes and cigars who inhaled moderately

or deeply were categorized as smokers or exsmokers, according to their current smoking status. The index pack-years, however, was derived entirely from cigarette smoking experience.

Indexes of Exposure

Included in the occupational history were estimates of total and autopsy cases personally embalmed by year for each year in which the person had embalmed at all and the usual number of hours spent embalming a normal or autopsy case. Exposure indexes were constructed either from estimated lifetime total of intact or autopsied bodies embalmed or from "total body index" (TBI), a weighted score that took into account both intact and autopsied bodies. Since on average the length of time required to embalm an autopsy case (2.8 h) was reported to be 1.5 times that of a normal case (1.9 h) and since exposures encountered in embalming an autopsy case were approximately three times as intense, TBI was defined as the lifetime total of intact bodies embalmed plus 4.5 times the total of autopsied bodies.

The validity of questionnaire data on numbers of bodies embalmed was assessed to the extent possible by reviewing funeral home records for a small number of study participants. An exact count of total bodies embalmed in given years was obtained for a few individuals. The ratio of total bodies reported on the questionnaire to the exact count is given for five persons for whom funeral home records were felt to be complete over several years, as follows: 0.8, 1.1, 1.9, 1.3, and 3.1.

Pulmonary Function Examinations

Subjects were called for pulmonary function testing by month of birth and were assigned in sequence to one of eight test stations in order to randomize inequities of machines and operators. Lung function determinations were conducted with Collins 13.5-liter respirometers. Three FVC maneuvers were obtained from each subject. The start of timing for FEV_1 measurements was determined by back extrapolation. The largest of three FVC and FEV_1 volumes was recorded, even if the two values did not come from the same curve. The FVC maneuver with the greatest sum of FVC and FEV_1 was used to compute FEF rates. Gas volumes were converted from room temperature, water vapor saturated, to 37° saturated; and corrections were made to account for deformations of the bell in the case of spirometers with metal bells. Predicted values for normal individuals of the same age and height were taken from nomograms developed by Morris (9). Conventional criteria cited by Miller were used to define pulmonary function abnormalities (10).

Analysis

Multiple regression methods were employed to investigate associations between pulmonary function variables and age, height, pack-years, years since last smoked, and exposure indexes. Significant associations between exposure indexes and pulmonary function variables were scrutinized by examining the distribution of the variables for normality and performing regressions on ranked data, when assumptions of normality were violated.

A stepwise multiple logistic regression analysis was used to assess the ability of TBI, age, and pack-years to predict respiratory symptoms. Variables were selected to include in the model according to the magnitude of their Q statistic: the variable with the highest value of Q entered the model first, and so forth (11). Those that did not contribute to the model significantly ($p < 0.10$) were deleted.

A matched pairs analysis was also performed to evaluate the relationship of exposure to respiratory symptoms and pulmonary function. Study participants were divided into high and low exposure groups stratified by smoking status. Within smoking categories, individuals were ranked by TBI. Starting from the subject with the lowest TBI and proceeding by rank order, a match was selected whose TBI was the highest of persons within 3 years of the subject's age. Of 37 nonsmokers, 30 exsmokers, and 32 smokers who completed questionnaires and pulmonary function tests, matched pairs were found for 32, 26, and 28, respectively. Levels of significance for symptoms were obtained by McNemar's test (12). The matched t-test was used to evaluate differences in pulmonary function variables (13).

RESULTS

Pulmonary Function

The prevalence of pulmonary function abnormalities and chronic bronchitis and the means of pulmonary function variables that can be compared to the normal population surveyed by Morris (9) are presented in Table 1 by smoking history. The expected variation with smoking is observed. Two nonsmokers, one exsmoker, and three smokers completed questionnaires but did not participate in lung function testing. These were not included in the computations for Table 1, and none had chronic bronchitis.

In Table 2 the prevalence of spirometric abnormalities among nonsmokers is noted for morticians, healthy Oregon men who had "never worked in a polluted atmosphere for any extended period (9, 10, 14), and a representative cross section of Michigan men (14). Morticians were older, but the reported prevalence of spirometric abnormalities in the three groups is similar.

Eight current or former pipe or cigar smokers were classified as nonsmokers. Five had no pulmonary function abnormalities, three had abnormal FEF_{25-75},

Table 1 Lung Function and Chronic Bronchitis Among White Male Morticians by Smoking History

	Nonsmokers (n = 37)		Exsmokers (n = 30)		Smokers (n = 32)	
	Mean (SE)*	% Abnormal†	Mean (SE)*	% Abnormal†	Mean (SE)*	% Abnormal†
FVC, % predicted	99.3 (1.7)	0.0	96.5 (2.3)	3.3	90.9 (3.1)	25.0
FEV_1, % predicted	107.2 (2.3)	5.4	101.7 (3.1)	6.7	91.9 (3.4)	15.6
FEV_1/FVC	0.803‡ (0.01)	5.4	0.778 (0.02)	13.3	0.750 (0.01)	25.0
FEF_{25-75}, % predicted	111.2 (5.7)	21.6	95.4 (6.8)	23.3	77.8 (5.2)	40.6
Chronic bronchitis, %	8.1		16.7		34.4	
Age (yr)	49.1 (2.3)		49.6 (2.5)		48.4 (2.4)	
Height (in)	68.3 (0.4)		68.8 (0.4)		69.1 (0.4)	
Smoking (pack-yr)	0.0		22.9 (3.5)		31.5 (3.4)	

*SE = standard error.

† Abnormal for FVC and FEV_1 is ≤ 79; for FEV_1/FVC is ≤ 0.74 for age < 30 yr, 0.69 for age 30-59 yr, 0.64 for age ≥ 60 yr; for FEF_{25-75}, is ≤ 74.

‡ The value of FEV_1/FVC in the Morris series (9) at the mean age and height of the mortician nonsmokers is 0.777.

Table 2 Prevalence of Spirometric Abnormalities Among Nonsmoking West Virginia Morticians and White Males from Oregon and Michigan

		% Abnormal		
Pulmonary function abnormalities		Morticians ($n = 37$)	Oregon men[*] ($n = 507$)	Michigan men[*] ($n = 154$)
FVC	$\leqslant 79\%$ predicted[†]	0.0	5.5	6
FEV_1	$\leqslant 79\%$ predicted[†]	5.4	5.9	1
FEF_{25-75}	$\leqslant 74\%$ predicted[†]	21.6	17.8	16
FEV_1/FVC	$\leqslant 0.74$, age < 30 yr $\leqslant 0.69$, age 30–59 yr $\leqslant 0.64$, age $\geqslant 60$ yr	5.4	19.7	
Age (yr)		49.1	37.9	42.1

[*]Values for Oregon men were derived from the population examined by Morris (9) and reported by Miller (10, 14). Values for Michigan men were taken from Miller (14).
[†]Predicted values were obtained from Morris (9).

and one had abnormal FEV_1. None had chronic bronchitis. No individuals were classified as exsmokers or smokers solely on the basis of pipe or cigar smoking.

Regression equations using TBI as the exposure index are given for the total group and for nonsmokers in Tables 3 and 4. Coefficients for age, height, and smoking (pack-years) have the expected sign, and the association of these variables with appropriate measures of pulmonary function frequently attains statistical significance. Within the total group, TBI was significantly related to

Table 3 Regression Equations to Predict Pulmonary Function: Total Group

Variable	N	Intercept	Age (yr)	Height (in)	Smoking (pack-yr)	Exposure (TBI)	Multiple R
FVC	99	−4.666	−0.026[§]	+0.154[§]	−0.012[‡]	+0.0000104	0.544
FEV_1	99	−1.944	−0.030[§]	+0.105[§]	−0.014[§]	−0.0000060	0.614
FEV_1/FVC	99	+115.336	−0.212[‡]	−0.349	−0.130[‡]	−0.0002583	0.385
FEF_{0-1000} ml	99	−1.729	−0.067[†]	+0.194	−0.022	−0.0000469	0.217
FEF_{25-75}	99	+2.057	−0.049[§]	+0.063	−0.023[‡]	−0.0000421	0.464
FEF_{25}	99	−0.279	−0.039[*]	+0.141	−0.032[†]	−0.0000381	0.203
FEF_{50}	99	−0.679	−0.042[‡]	+0.112	−0.029[‡]	−0.0000525	0.371
FEF_{75}	99	+2.914	−0.033[§]	+0.005	−0.111[‡]	−0.0000220	0.494
$FEF_{200-1200}$ ml	98	−0.866	−0.048[†]	+0.147	−0.020	−0.0000195	0.178
FEF_{50}/FVC	99	+164.509	−0.378	−0.570	−0.494[†]	−0.0014750[*]	0.196

[*]$p \leqslant 0.05$, one-tailed.
[†]$p \leqslant 0.01$, one-tailed.
[‡]$p \leqslant 0.001$, one-tailed.
[§]$p \leqslant 0.0001$, one-tailed.

Table 4 Regresssion Equations to Predict Pulmonary Function: Nonsmokers

Variable	N	Intercept	Age (yr)	Height (in)	Exposure (TBI)	Multiple R
FVC	37	−8.319	−0.027 [§]	+0.208 [§]	+0.0000081	0.727
FEV_1	37	−6.554	−0.029 [§]	+0.172 [§]	−0.0000109	0.692
FEV_1/FVC	37	+78.778	−0.153 [*]	+0.149	−0.0003439 [*]	0.221
FEF_{0-1000} ml	37	−2.859	−0.057 [*]	+0.202	−0.0000820	0.221
FEF_{25-75}	37	−4.536	−0.048 [†]	+0.162 [*]	−0.0000607	0.399
FEF_{25}	37	−1.986	−0.057 [*]	+0.179	−0.0000347	0.173
FEF_{50}	37	−6.656	−0.043 [*]	+0.201 [*]	−0.0000497	0.287
FEF_{75}	37	+4.276	−0.040 [‡]	−0.008	−0.0000345	0.374
$FEF_{200-1200}$ ml	36	−13.687	−0.019	+0.312 [*]	−0.0000535	0.196
FEF_{50}/FVC	37	+99.590	−0.325	+0.338	−0.0014302	0.081

[*]$p \leqslant 0.05$, one-tailed.
[†]$p \leqslant 0.01$, one-tailed.
[‡]$p \leqslant 0.001$, one tailed.
[§]$p \leqslant 0.0001$, one tailed.

1 of 10 pulmonary function variables: FEF_{50}/FVC ($p = 0.04$, one-tailed). FEF_{50}/FVC, however, was not normally distributed; moreover, the association between it and TBI was found not to be significant ($p = 0.25$, one-tailed) in a regression using ranked data. In order to discern effects that may have been masked by aggregating persons of different smoking habits, regression equations were derived separately for each smoking group. This series of 30 analyses produced only one significant association with TBI: FEV_1/FVC for nonsmokers ($p = 0.04$, one-tailed). Here again, analysis of ranked data indicated that the association was unstable ($p = 0.12$, one-tailed).

Since exposures during the embalming of autopsied bodies are considerably more intense, perhaps only these are of biological significance. If such were true, TBI might have been an inappropriate exposure index to relate to health since it was derived from intact as well as autopsied bodies. Additional analyses of the entire group and of individual smoking categories, therefore, were performed, replacing TBI with total autopsied bodies embalmed. Only 1 of these 40 analyses—FEF_{50}/FVC for the entire group—was significantly associated with autopsied bodies ($p = 0.05$, one-tailed). As indicated above, the variable FEF_{50}/FVC was not normally distributed; in addition, an analysis by rank determined that the relationship with autopsied bodies was not significant ($p = 0.24$, one-tailed).

Characteristics of smoking-specific, age-matched exposure groups are displayed in Table 5. Smokers with high exposure were shorter and had smoked more than their low exposure counterparts. These differences of themselves would tend to increase respiratory symptoms and decrease pulmonary function in the high exposure group. On the other hand, exsmokers with high exposure were taller, had smoked less, and had abstained longer than their counterparts, which would decrease respiratory symptoms and increase pulmonary function in the high exposure group. Seven current or former pipe or cigar smokers were included in the matched groups of nonsmokers: three were in the group with low exposure and four were in the group with high exposure. None of those included in the low exposure group had pulmonary function abnormalities, but two pipe or cigar smokers with pulmonary function abnormalities were assigned to the high exposure group: one had abnormal FEV_1, and both had abnormal FEF_{25-75}.

The 10 pulmonary function variables for which regression equations were derived and percent predicted FVC-FEV_1-FEF_{25-75}-$FEF_{200-1200}$ were compared between age-matched high and low exposure groups within smoking categories. Significant differences were detected for 2 of the 42 comparisons—FEF_{25-75} and FEF_{75} among nonsmokers—both at the one-tailed $0.025 > p < 0.050$ level.

Respiratory Symptoms

Figures 1–3 describe cough, phlegm, and wheeze among matched exposure groups by smoking category. Other than the expected effects of smoking, no consistent pattern associated with high exposure was discerned; moreover, none of the differences observed was statistically significant. The association of chronic bronchitis with smoking and its lack of association with high exposure is vividly depicted in Fig. 4.

A stepwise multiple logistic regression analysis was performed to determine whether TBI, age, and pack-years were useful indicators of symptoms described in Figs. 1–4. While pack-years was significantly associated with most symptoms, TBI and age were not significant factors for any.

Exposure

For each year in which they embalmed, on the average, morticians embalmed approximately 75 bodies, of which 15 were autopsy cases (for an approximate

Table 5   Characteristics of Matched Groups: Means

	Nonsmokers		Exsmokers		Smokers	
Exposure	Low	High	Low	High	Low	High
N	16	16	13	13	14	14
Age (yr)	47.7	48.3	50.3	51.4	48.6	49.0
Height (in)	68.8	68.4	68.0	69.2	69.5	68.6
Cigarettes (pack-yr)			26.0	21.3	29.3	36.2
Quit smoking (yr)			11.9	14.6		
TBI	988.9	6269.4	1629.2	4857.3	2202.5	5697.7

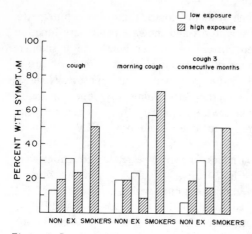

Figure 1 Percent with cough among smoking-specific, high and low exposure groups of white male morticians matched for age. Non = nonsmokers; Ex = exsmokers; Smokers = current smokers. "Cough" refers to usually having a cough.

TBI of 130). Three of the ninety-nine morticians from whom complete data were obtained had never embalmed. All were nonsmokers, but one had a pulmonary function abnormality—a mild reduction of FEV_{25-75} (74 percent of expected, whereas $\leqslant 74$ percent was considered abnormal). At the other extreme of embalming experience, one mortician had prepared over 10,000 bodies, including more than 1800 autopsy cases, another had embalmed over 7000 bodies, including more than 6200 autopsy cases. The former had chronic bronchitis and a mild reduction of FEF_{25-75} (73 percent of expected). Both were non-smokers.

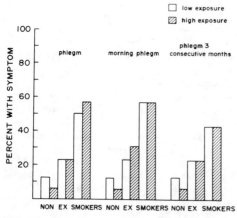

Figure 2 Percent with phlegm among smoking-specific, high and low exposure groups of white male morticians matched for age. Non = nonsmokers; Ex = exsmokers; Smokers = current smokers. "Phlegm" refers to usually bringing up phlegm.

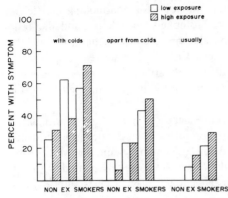

Figure 3 Percent with wheeze among smoking-specific, high and low exposure groups of white male morticians matched for age. Non = nonsmokers; Ex = exsmokers; Smokers = current smokers. "Apart from colds" refers to occasional wheezing apart from colds. No nonsmokers reported usually having a wheeze.

DISCUSSION

The effects of occupational exposure on the respiratory health of morticians were assessed in three ways: First of all, pulmonary function means and abnormalities for the entire study group were compared with those of normal populations from Oregon and Michigan. Second, age-matched high and low exposure subgroups within smoking categories were examined to identify correlations between exposure, respiratory symptoms, and lung function decrements. Third, multiple regression analyses of the association of exposure indexes with respiratory symptoms and pulmonary function were performed. Inaccuracies of exposure indexes are minimized in the matched group comparisons and avoided altogether in comparisons of the entire study group with normal populations.

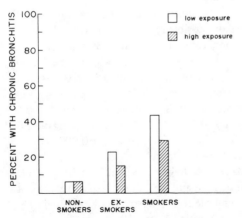

Figure 4 Percent with chronic bronchitis among smoking-specific, high and low exposure groups of white male morticians matched for age.

Mean values of pulmonary function variables for nonsmoking morticians were similar to or greater than those of nonsmokers from Oregon and Michigan. The prevalence of pulmonary function abnormalities was similar, despite the fact that the morticians were older. Mean FEV_1 was expected to be greater than that of the Oregon population since the start of timing was determined by back extrapolation, whereas the Oregon values were computed using the Kory method (15).

Age, pulmonary function, and the prevalence of spirometric abnormalities for exsmokers were comparable among morticians and Michigan residents. Morticians who smoked, however, had lower pulmonary function values, probably the result of differences in pack-years of exposure associated with age. (The mean age of mortician and Michigan smokers was 48 years and 38 years, respectively.)

Comparison of smoking-specific, age-matched exposure groups revealed no correlation between high exposure and the prevalence of cough, phlegm, wheeze, or chronic bronchitis. Pulmonary function differences of borderline significance were detected for 2 of 42 comparisons—FEF_{25-75} and FEF_{75} among nonsmokers. These may have been due to smoking rather than exposure since the two pipe or cigar smokers with abnormal FEF_{25-75} were included in the high exposure group.

Eighty multiple regressions revealed only three significant associations between exposure and decrements in pulmonary function. Not one of these proved robust when regressions were performed using ranked data.

With so many tests being performed, significance testing is considered to be more useful to indicate possible rather than probable correlations, especially when p-values are close to the 0.05 level. Because of interdependence, had one expiratory flow rate been substantially affected it was felt that others would have been as well. In view of these matters, the few significant associations observed were interpreted as chance findings.

Although regression coefficients for exposure were not statistically significant, they provide the only estimates of possible magnitude of effect. Using the regression coefficients of Table 3, one can estimate the potential effect of lifetime exposure. Suppose that a mortician begins to embalm at age 20 and retires at age 65, having accumulated 45 years of exposure. Suppose also that his average annual exposure is the group average TBI of 130, giving a career total of 5850. Since the regression coefficient is positive, FVC would not be affected adversely. Exposure would decrease FEV_1 and FEF_{25-75} by the product of the regression coefficients for TBI (-0.000006 and -0.000042) and 5850, indicating losses of 0.035 and 0.246 liter, respectively, over a lifetime. Assuming that the mortician were of average height (68.7 inches), at age 65 his expected FEV_1 and FEF_{25-75} would be 2.98 and 2.82 liters, respectively. By age 65, therefore, he might have lost the equivalent of 1.2 percent of expected FEV_1 and 8.7 percent of expected FEF_{25-75}, potential losses of little consequence.

Had group lung function decreased substantially with exposure, would a statistically significant difference have been recognized in multiple regression analyses? If one uses conventional criteria for abnormality to define functional deficients of clinical importance, by age 65 the FVC and FEV_1 would be reduced by 20 percent (0.86 and 0.60 liter) and FEF_{25-75} would be reduced 25 percent (0.71 liter). The regression coefficients needed for career TBI exposure to result in losses of such magnitude are -0.000154, -0.000103, and -0.000121 for FVC, FEV_1, and FEF_{25-75}, respectively. The magnitude of regression coefficients that could have been detected with a power of 0.90 at the $p = 0.05$ level is given by the product of the standard error of TBI and -2.927 standard deviation units: -0.000049, -0.000043, and -0.000090 for FVC, FEV_1, and FEF_{25-75}, respectively. Since these are all smaller than regression coefficients needed to produce important functional deficits, substantial decreases in group lung function would indeed have been recognized.

While it is possible that significant quantities of formaldehyde might be carried deep into the lung on particles, its high aqueous solubility and chemical reactivity suggest that absorption of the gas during nose breathing would occur primarily in the upper respiratory tract. At concentrations of up to 25 ppm administered to healthy human volunteers, virtually all sulfur dioxide, another highly water-soluble gas, is removed by the nose (16, 17). Absorption of formaldehyde gas administered at 15 ppm to Fischer-344 rats, which are obligatory nose breathers, appears to predominate in the upper respiratory tract (see Heck et al., Chap. 4, this volume). It is not known whether inhalation by mouth might be associated with less efficient absorption in the upper airways.

A survey of embalmers employed in funeral homes in the vicinity of Los Angeles noted an apparent increased prevalence of acute or chronic bronchitis among nonsmokers (3). No control group was used, however; and asymptomatic men had worked longer than those with symptoms. a fact that may contradict the purported association with occupational exposure.

An investigation of workers producing acrylic-wool filters impregnated with phenol-formaldehyde resin found an excess of chronic cough and phlegm and chronic airway obstruction evidenced by reduced Monday preshift FEV_1/FVC and FEF_{50}/FVC (2). These results, which are at variance with the present study, may perhaps be attributed to greater duration or intensity of exposure of the acrylic-wool filter workers; to the presence of phenol, additives, and resin and fiber decomposition products; or to residual acute effects, which would be less likely to occur among morticians due to the intermittent nature of their exposure. (The average mortician who embalms does so only a few hours a week, whereas many filter workers were reported to have worked on Saturday before the Monday measurements.) An hypothesis of residual acute effects would help to reconcile differences in observations made among filter and rubber workers (1).

In summary, pulmonary function of West Virginia morticians compared favorably to that of residential populations in Oregon and Michigan. Among morticians, high exposure was linked neither to chronic bronchitis nor to pulmonary function deficits. The results suggest that intermittent exposure to low levels of formaldehyde gas over the long term exerts no meaningful chronic effect on respiratory health.

REFERENCES

1 Gamble, JF, AJ McMichael, T Williams, and M Battigelli. Respiratory function and symptoms: An environmental-epidemiological study of rubber workers exposed to a phenol-formaldehyde type resin. *Am. Ind. Hyg. Assoc. J.* 37:499–513 (1976).

2 Schoenberg, JB, and CA Mitchell. Airway disease caused by phenolic (phenol-formaldehyde) resin exposure. *Arch. Environ. Health* 30:574–577 (1975).

3 Plunkett, ER, and T Barbela. Are embalmers at risk? *Am. Ind. Hyg. Assoc. J.* 38:61–62 (1977).

4 Hendrick, DJ, and DJ Lane. Occupational formalin asthma. *Br. J. Ind. Med.* 34:11–18 (1977).

5 Williams, TR, RJ Levine, and PB Blunden. Chemical exposure of embalmers. Occupational Health Studies Group, University of North Carolina at Chapel Hill (1980).

6 Strub, CG, and LG Frederick. *The Principles and Practice of Embalming*, 4th ed. Dallas: LGD Frederick (1967).

7 Mayer, JS. *Restorative Art*, 7th ed. Cincinnati: Paula Publishing Co. (1974).

8 Ferris, BG. Epidemiology standardization project. *Am. Rev. Resp. Dis.* 118:1–220 (1978).

9 Morris, JF, A Koski, and LC Johnson. Spirometric standards for healthy nonsmoking adults. *Am. Rev. Resp. Dis.* 103:57–67 (1971).

10 Miller, A, JC Thornton, H Smith, Jr., and JF Morris. Spirometric "abnormality" in a normal male reference population. *Am. J. Ind. Med.* 1:55–68 (1980).

11 Harrell, F. The logist procedure. In *SAS Supplemental Library User's Guide*, edited by PS Reinhardt, pp. 83–102. Cary, NC: SAS Institute Inc. (1980).

12 Fleiss, JL. *Statistical Methods for Rates and Proportions*, p. 74. New York: Wiley & Sons (1973).

13 Armitage, P. *Statistical Methods in Medical Research*, pp. 116–118. New York: Wiley & Sons (1971).

14 Miller, A, and JC Thornton. The interpretation of spirometric measurements in epidemiologic surveys. *Environ. Res.* 23:444–468 (1980).

15 Kory, RC, R Callahan, and HG Boren. The Veterans Administration-Army Cooperative Study of Pulmonary Function. I. Clinical spirometry in normal men. *Am. J. Med.* 30:243–258 (1961).

16 Speizer, FE, and NR Frank. The uptake and release of SO_2 by the human nose. *Arch. Environ. Health* 12:725–728 (1966).

17 Andersen, I, GR Lundqvist, PL Jensen, and DF Proctor. Human response to controlled levels of sulfur dioxide. *Arch. Environ. Health* 28:31–39 (1974).

Chapter 20

Proportionate Mortality among New York Embalmers

Judy Walrath

Joseph F. Fraumeni, Jr.

Since the turn of the century, formaldehyde has been the main preservative in commercial embalming fluids (1). In an industrial hygiene survey of a mortuary science college conducted by the National Institute for Occupational Safety and Health, the airborne formaldehyde concentrations exceeded 3.0 ppm in two samples when the ventilation system was closed and dropped to 0.20-0.91 ppm when the system was operating (2). A survey of six funeral homes revealed airborne formaldehyde levels ranging from 0.09-5.26 ppm, with average concentrations of 0.25-1.39 ppm. Paraformaldehyde particles with formaldehyde vapors were small enough to be deposited in the lungs (3).

In humans, formaldehyde irritates the eyes, skin, and respiratory system, but the chronic effects of exposure are unknown (4). Serious concerns were raised by preliminary results of an animal inhalation study by the Chemical Industry Institute of Toxicology, in which rats exposed to 15 ppm formaldehyde vapors developed a marked excess of nasal cancers (see Kerns et al., Chap. 11, this volume).

This preliminary study investigates whether embalmers, compared with the general population, have a greater proportion of cancer deaths that might be associated with exposure to formaldehyde.

METHODS

The licensure divisions of state health departments in several states were surveyed to determine whether records existed from which a cohort of licensed embalmers could be established. All states contacted lacked the records necessary for a retrospective cohort study. New York State, however, had maintained sufficient information on deceased embalmers so that it was possible to examine the proportion of deaths due to specific causes.

The study group consists of deceased embalmers licensed to practice embalming in New York State between 1902 and 1979. Names of decedents from 1925–1979 were obtained from records of the Bureau of Funeral Directing and Embalming in the New York State Department of Health. The Bureau offers two types of licenses: a) license for embalming and b) license for funeral directing (i.e., the general management of a funeral home). A qualified individual may apply for one license or two but cannot combine the tasks of embalming and funeral directing without having both licenses. Since exposure to formaldehyde was the variable of interest, deaths among persons who held only a funeral director's license were not included. Study subjects were identified through two data sources at the Bureau. The first was registration files of embalmers who were known to be deceased during recent years and whose files were kept at the Bureau. Each decedent's name, date of birth, first and last years of licensure, last known address, and date of death were abstracted from these files. The second source of data was a ledger in which names of all registered embalmers were entered at time of initial application. When the office was notified of a death, this fact was recorded in the ledger. For each decedent listed, the name, first year of licensure, residence, and date of death were abstracted. Forty-two percent of the study group was ascertained from registration files and 58 percent from the ledger.

Death certificates were requested for 1376 embalmers from the appropriate state vital statistics offices. At the present time, death certificates have been found for 92 percent of the decedents; 131 of these have not yet been received. Underlying cause of death was coded by a nosologist using the rules in effect at time of death and the rubrics of the 8th Revision of the International Classification of Diseases, Adapted (ICDA) (5). The deaths observed among the embalmers were compared to expected numbers computed by applying the age-, race-, sex-, and calendar year-specific proportions of deaths for each cause among the U.S. general population to the total number of deaths in the study group by 5-year age and time periods (6). Differences between observed and expected numbers of deaths for each cause were summarized in the form of the proportionate mortality ratio (PMR), which is the ratio of the number of deaths observed to that expected multiplied by 100. The statistical significance of each ratio was tested by a chi^2 test with one degree of freedom (7). PMRs and chi^2 values were not presented in tables for the total study group when both the observed and expected numbers of deaths were less than five. Proportionate cancer mortality ratios (PCMRs) were also computed utilizing the total number of cancer deaths as the denominator for calculating the expected number of deaths for each individual cancer site (6).

It was not possible to measure length of employment or length of licensure in the analysis since year of last license was not available for decedents who were listed only in the ledger. As an alternative, length of time from first license to death was considered in the analysis.

RESULTS

Table 1 shows the race and sex distribution of the 1106 deceased embalmers. There were 1010 (91.3 percent) white males and 67 (6.1 percent) nonwhite males in the study group. The 24 males (2.2 percent) whose race was unknown and the 5 females were excluded from the analysis. (Their mortality patterns were examined separately and were not unusual.)

Table 2 shows the age distribution of the male embalmers at time of death by year of death. Fifty percent of the embalmers died before age 65. This relatively young age distribution is probably related to the fact that the Bureau is usually notified of the death of a registrant when a license renewal form is returned to the office by a surviving relative. Ninety-five percent of the embalmers for whom year of last license was known died within one year of license termination. Embalmer was listed as the usual occupation on 90 percent of the death certificates. The median birth year for those in the study group was 1901 and the median year of initial license was 1931.

Table 3 shows the numbers of deaths observed and expected and the PMRs for five major causes of death among white males. The excess of deaths from all malignant neoplasms (PMR = 108) was not statistically significant. An elevated ratio of deaths due to diseases of the circulatory system was largely attributable to a significant excess of arteriosclerotic heart disease (PMR = 111). There was a deficit of diseases of the respiratory system, with deaths from pneumonia being significantly lower than expected. Among diseases of the digestive system, there was an elevated mortality due to cirrhosis of the liver. The PMR was significantly low for accidental deaths but slightly elevated for suicides.

The number of deaths observed among nonwhites was small (Table 4). As

Table 1 Distribution of Deceased Embalmers by Sex and Race

	White	Nonwhite	Unknown race	Total
Male	1010	67	24	1101
Female	4	1	0	5
Total	1014	68	24	1106

Table 2 Distribution of Deaths Among Male Embalmers by Age at Death and Year of Death, All Races Combined

Age at death	1925–54	1955–59	1960–64	1965–69	1970–74	1975–79	All calendar years	%
20–39	7	4	6	5	3	3	28	2.6
40–44	7	12	5	7	6	8	45	4.2
45–49	12	9	9	6	10	9	55	5.1
50–54	13	25	8	17	20	13	96	8.9
55–59	19	20	14	42	30	25	150	13.9
60–64	20	25	21	24	38	31	159	14.8
65–69	18	26	20	42	30	43	179	16.6
70–74	13	9	23	37	36	36	154	14.3
75–79	7	9	10	25	34	24	109	10.1
80+	5	12	6	10	36	33	102	9.5
All age groups	121	151	122	215	243	225	1077	
%	11.2	14.0	11.3	20.0	22.6	20.9		

Table 3 Numbers of Deaths and Proportionate Mortality Ratios (PMR) for White Male Embalmers by Major Cause of Death

Cause of death (ICDA Code, 8th Revision)	Number of deaths		PMR	Chi² value
	Observed	Expected		
Malignant neoplasms (140–209)	210	193.7	108	1.71
Circulatory system (390–458)	587	558.6	105	3.34
Arteriosclerotic heart disease (410–414)	428	384.8	111	7.97[†]
Cerebrovascular diseases (430–438)	87	80.1	109	0.66
Respiratory system (460–519)	47	60.0	78	2.99
Pneumonia (480–486)	13	22.4	58	4.08[*]
Emphysema (492)	11	15.9	69	1.59
Digestive system (520–577)	53	47.0	113	0.84
Gastric and duodenal ulcers (531, 532)	8	7.9	102	0.00
Cirrhosis of liver (571)	29	22.5	129	1.94
External causes (800–999)	50	74.4	67	9.67[†]
Accidents (800–949)	27	50.4	54	12.36[†]
Suicide (950–959)	21	18.5	114	0.36
All other causes	63	76.3	83	
All causes of death	1010	1010.0	100	

[*] $p < 0.05$.
[†] $p < 0.005$.

Table 4   Numbers of Deaths and Proportionate Mortality Ratios (PMR) for
Nonwhite Male Embalmers by Major Cause of Death

Cause of death (ICDA Code, 8th Revision)	Number of deaths		PMR	Chi^2 value
	Observed	Expected		
Malignant neoplasms (140–209)	18	12.2	148	3.49
Circulatory system (390–458)	38	35.8	106	0.29
Arteriosclerotic heart disease (410–414)	27	17.9	151	6.46[*]
Cerebrovascular diseases (430–438)	3	8.5	35	4.08[*]
Respiratory system (460–519)	0	4.0		4.29[*]
Pneumonia (480–486)	0	2.2		
Emphysema (492)	0	0.5		
Digestive system (520–577)	4	2.3		
Gastric and duodenal ulcers (531, 532)	1	0.4		
Cirrhosis of liver (571)	1	0.9		
External causes (800–999)	1	4.2		
Accidents (800–949)	1	2.8		
Suicide (950–959)	0	0.2		
All other causes	6	8.5	71	
All causes of death	67	67.0	100	

[*]$p < 0.05.$

in white males, there was a nonsignificant excess of malignant neoplasms (PMR = 148) and a significant excess of arteriosclerotic heart disease (PMR = 151). Deaths due to respiratory diseases and external causes were lower than expected.

The distribution of malignant neoplasms by primary site among white males is shown in Table 5. The number of deaths from cancer of the buccal cavity and pharynx approximated the expected value and did not include any deaths from nasopharyngeal cancer. Mortality from cancers of the digestive tract was close to that expected, but there was an excess of colon cancer and a deficit of rectal cancer. Respiratory cancer mortality was not unusual. No deaths were attributed to nasal cancer, although the expected value was only 0.7. There was a significant excess mortality from skin cancer. Four of the eight cases were malignant melanoma (PMR = 231), three were squamous cell carcinoma, and one was unspecified. There was a slight excess of kidney cancer (8 deaths versus 4.7 expected), cancer of the brain and central nervous system (8 deaths versus 5.1 expected), and leukemia (10 deaths versus 7.6 expected). The cell types of leukemia were as follows: five myelogenous (four acute and one NOS), one acute monocytic, three lymphatic (two chronic and one NOS), and one acute leukemia NOS.

Proportionate cancer mortality ratios among white males were low for cancers of the stomach (PCMR = 72), rectum (PCMR = 23), and prostate (PCMR = 79). Elevated PCMRs were seen for cancers of the lung (PCMR = 107),

Table 5   Numbers of Deaths Due to Malignant Neoplasms and Proportionate Mortality Ratios (PMR) for White Male Emblamers

Cause of death (ICDA Code, 8th Revision)	Number of deaths		PMR	Chi² value
	Observed	Expected		
All malignant neoplasms (140–209)	210	193.7	108	1.71
Buccal cavity and pharynx (140–149)	8	6.4	126	0.43
Digestive organs and peritoneum (150–159)	59	58.4	101	0.01
Esophagus (150)	4	4.7		
Stomach (151)	11	12.2	90	0.12
Colon (153)	25	18.0	140	2.78
Rectum (154)	2	7.0	29	3.59
Liver and gallbladder (155, 156)	4	4.3		
Pancreas (157)	12	11.0	110	0.10
Respiratory system (160–163)	63	61.9	102	0.02
Larynx (161)	2	3.0		
Lung and pleura (162, 163)	61	58.2	105	0.14
Skin (172, 173)	8	3.2	253	7.40*
Prostate (185)	13	14.5	89	0.16
Bladder (188)	6	6.5	92	0.04
Kidney (189)	8	4.7	170	2.26
Brain and central nervous system (191, 192)	8	5.1	157	1.68
Lymphatic and hematopoietic system (200–209)	21	18.2	115	0.44
Lymphosarcoma and reticulosarcoma (200)	4	4.2		
Hodgkin's disease (201)	2	2.0		
Other lymphatic cancers (202, 203)	5	4.2	118	0.14
Leukemia (204–207)	10	7.6	132	0.80
Other cancers	16	14.8	108	

*$p < 0.01$.

skin (PCMR = 206), kidney (PCMR = 157), and brain (PCMR = 136). This pattern is consistent with the PMR analysis except the PCMR was lower than expected for cancers of the stomach and prostate.

The distribution of malignant neoplasms among nonwhite males is presented in Table 6, although the numbers of deaths are small.

Mortality for selected cancer sites among white males was examined by latency period, defined as length of time from first license to death (Table 7). Since almost all of the embalmers in this study were licensed at time of death, latency period is assumed to be synonymous with length of licensure. Although the numbers involved are small, mortality from skin cancer was significantly greater than expected (PMR = 354) among those licensed for 35+ years. On the other hand, the elevated PMR for kidney cancer was limited to embalmers licensed under 35 years. No unusual patterns were seen for cancers of the brain and lymphatic-hematopoietic system.

Mortality patterns by age at first license are shown in Table 8. Excess

Table 6 Numbers of Deaths Due to Malignant Neoplasms and Proportionate Mortality Ratios (PMR) for Nonwhite Male Emblamers

Cause of death (ICDA Code, 8th Revision)	Number of deaths		PMR	Chi^2 value
	Observed	Expected		
All malignant neoplasms (140–209)	18	12.2	148	3.49
Buccal cavity and pharynx (140–149)	0	0.4		
Digestive organs and peritoneum (150–159)	5	4.1	123	0.23
Respiratory system (160–163)	5	3.3	150	0.89
Larynx (161)	2	0.2		
Lung (162, 163)	3	3.1		
Skin (172, 173)	0	0.1		
Prostate (185)	3	1.9		
Bladder (188)	1	0.3		
Kidney (189)	1	0.2		
Brain and central nervous system (191, 192)	0	0.1		
Lymphatic and hematopoietic system (200–209)	2	0.8		
Other cancers	1	1.0		

mortality due to skin and brain cancers was seen among those who began to practice embalming at age 30 or later.

Cancer mortality was examined separately for persons licensed only as embalmers and for those who held licenses for both embalming and funeral directing. We assumed that persons licensed only as embalmers accumulated a longer average exposure to formaldehyde than embalmers who were also

Table 7 Observed Numbers of Deaths with Proportionate Mortality Ratios for White Male Embalmers by Length of Time From First License to Death

Underlying cause of death category (8th Revision, ICDA)	Length of time from first license to death					
	< 35 yr ($n = 551$)			35+ yr ($n = 459$)		
	O	PMR	Chi^2	O	PMR	Chi^2
All malignant neoplasms (140–209)	109	107	0.54	101	110	1.26
Respiratory system (160–163)	34	106	0.11	28	94	0.11
Skin (172, 173)	4	196	1.90	4	354	7.32[†]
Kidney (189)	6	223	4.09[*]	2	98	0.00
Brain and central nervous system (191, 192)	6	169	1.72	2	129	0.13
Lymphatic and hematopoietic tissue (200–209)	11	106	0.04	10	127	0.60

O = Observed number of deaths.
[*]$p < 0.05$.
[†]$p < 0.01$.

Table 8   Observed Numbers of Deaths with Proportionate Mortality Ratios for White Male Embalmers by Age at First License

Underlying cause of death category (8th Revision, ICDA)	Age at first license					
	Before age 30 ($n = 578$)			After age 30 ($n = 432$)		
	O	PMR	Chi^2	O	PMR	Chi^2
All malignant neoplasms (140–209)	130	113	2.29	80	102	0.05
Respiratory system (160–163)	42	105	0.12	21	95	0.05
Skin (172, 173)	3	151	0.52	5	424	12.41*
Kidney (189)	4	136	0.37	4	224	2.77
Brain and central nervous system (191, 192)	3	84	0.09	5	328	7.98*
Lymphatic and hematopoietic tissue (200–209)	15	133	1.23	6	87	0.12

O = Observed number of deaths.
*$p < 0.01$.

Table 9   Observed Numbers of Deaths with Proportionate Mortality Ratios for White Male Embalmers by License Type

Underlying cause of death category (8th Revision, ICDA)	Type of license					
	Embalmer only ($n = 528$)			Both embalmer and funeral director ($n = 482$)		
	O	PMR	Chi^2	O	PMR	Chi^2
All malignant neoplasms (140–209)	103	108	0.70	107	109	1.02
Respiratory system (160–163)	26	94	0.09	37	107	0.21
Skin (172, 173)	5	337	8.34†	3	178	1.04
Kidney (189)	6	256	5.75*	2	84	0.07
Brain and central nervous system (191, 192)	6	245	5.17*	2	76	0.15
Lymphatic and hematopoietic tissue (200–209)	8	91	0.07	13	138	1.40

O = Observed number of deaths.
*$p < 0.025$.
†$p < 0.005$.

funeral directors. Table 9 shows that mortality from skin, kidney, and brain cancers was significantly elevated for those who were licensed only as embalmers. No unusual mortality was observed among those who held both licenses.

DISCUSSION

Using the proportionate mortality approach, embalmers in this study experienced a slightly elevated mortality from cancer, a significant excess of arteriosclerotic heart disease, and a significant deficit of pneumonia and accidental deaths. Skin cancer mortality was significantly elevated, with the excess primarily among those licensed for more than 35 years and those who began employment at age 30 or later. Also elevated was the proportionate mortality from kidney and brain cancers. There was no excess mortality from cancers of the respiratory tract, including the nasal passages. This is noteworthy in view of reports that formaldehyde induces nasal cancer in rats following inhalation. In our study, the excesses of cancers of the skin, kidney, and brain suggest that the further investigation of the carcinogenic effects of formaldehyde should not be limited to the respiratory system. It must be borne in mind, however, that embalming fluids contain a mixture of other chemicals (e.g., tissue moisturizers, antiseptic solutions, dyes, and deodorizers) that are partly intended to offset the adverse reactions of formaldehyde (1, 8).

This study was intended as a preliminary investigation of the chronic effects of exposure to formaldehyde and has several deficiencies. Length-of-employment information was unavailable for those in the study group, and ascertainment of deaths among retirees was incomplete. Of special concern are the weaknesses of the proportionate mortality method, especially the uncertainty that an excess proportion of deaths from a specific cause reflects a real elevation of mortality or a deficit in the proportion of deaths from other causes. Despite these limitations, our findings suggest a need for cohort studies of embalmers and other workers exposed to formaldehyde to quantify the risks of various cancers in relation to job exposures.

REFERENCES

1 The Champion Company. *Expanding Encyclopedia of Mortuary Practice*, No. 365 (March, 1966).
2 National Institute for Occupational Safety and Health. *Health Hazard Evaluation Determination*, Report No. HE 79-146-670, Cincinnati, OH: NIOSH (March, 1980).
3 Kerfoot, EJ, and TF Mooney, Jr. Formaldehyde and paraformaldehyde study in funeral homes. *Am. Ind. Hyg. Assoc. J.* 36:533–537 (1975).
4 National Institute for Occupational Safety and Health. *Criteria for a Recommended Standard . . . Occupational Exposure to Formaldehyde*, pp. 32–59.

DHEW (NIOSH) Publ. No. 77-126. Washington, DC: U.S. Government Printing Office (1976).

5 World Health Organization. *Manual of the International Statistical Classification of Diseases, Imjuries, and Causes of Death*, 8th rev. Geneva: World Health Organization (1967).

6 Monson, RR. Analysis of relative survival and proportional mortality. Comput. Biomed. Res. 7:325–332 (1974).

7 Mantel, N, and W Haenszel. Statistical aspects of the analysis of data from retrospective studies of disease. *J. Natl. Cancer Inst.* 22:719–748 (1959).

8 Kirk-Othmer: Embalming fluids. In *Encyclopedia of Chemical Technology*, Vol. 8, pp. 100–102. New York: Interscience (1965).

Chapter 21

Proportional Mortality among Chemical Workers Exposed to Formaldehyde

Gary M. Marsh

Formaldehyde, HCHO, and its derivatives are chemicals used in the manufacture, formulation, commercial distribution, and production of a variety of products, including textiles, foam insulation, resins, preservatives, and wood products. It has been estimated by NIOSH that 8000 employees may be at risk of exposure to the high concentrations of formaldehyde found in industrial synthesis, formulation, and distribution of concentrated products (1). The numerous uses of formaldehyde and its derivatives indicate, however, that a substantially larger population of employees may be at risk from intermittent exposures to products containing sources of formaldehyde or its congeners and derivatives (1).

Occupational exposures usually are the result of the release of free formaldehyde vapor into the air. Principal hazards that have been associated with human exposure to airborne formaldehyde are irritation of the respiratory tract, eyes, and skin (2). In 1979, the Chemical Industry Institute of Toxicology reported a high incidence of nasal carcinomas among rats exposed to formaldehyde vapor (Chapter 11).

Currently, there is little information available on the chronic toxicity and potential carcinogenicity related to human exposure to formaldehyde. This is a report on an epidemiologic investigation of workers exposed to formaldehyde at a large chemical-producing plant.

SPRINGFIELD STUDY

The Monsanto Company started using formaldehyde at its Springfield/Indian Orchard, Massachusetts plant in 1938. The chemical was used as a raw material in the manufacture of such products as phenolic resins, urea-formaldehyde resins, melamine-formaldehyde resins, hexamethylenetetramine, and resorcinol. Formaldehyde was first produced on a commercial scale at Springfield in 1948.

Exposure

Table 1, which summarizes the history of formaldehyde usage and manufacture at Springfield, shows that several of the original formaldehyde related processes were discontinued in the 1940s and 1950s. Clearly, there were some worker exposures, although no reliable quantitative information is available as to actual exposure levels. Exposure levels to formaldehyde vapor were probably higher, however, in the areas where formaldehyde was present in a liquid form (cast phenolics, resins, other Resinox, Resimines, and Formalin plant) compared to levels in areas where formaldehyde had already reacted with other chemicals to form solid or powdered materials (resins compounding and molding powder) (Henshaw J, personal communication, Nov. 13, 1980). In addition, due to major process changes, exposure levels in the late 1930s, 1940s, and 1950s are thought to have been substantially higher than in more recent years. Formaldehyde exposures that occurred in the Service Department-Waste Disposal are relatively

Table 1   Areas of the Monsanto Company Indian Orchard Plant Associated with Formaldehyde Exposure

Area	Time frame	Materials used
Cast phenolics	1938–1945	Formaldehyde, phenol, sodium solvents, lactic acid, lead molds, other
Resinox		Formaldehyde, phenol, methanol,
Resins	1939–Present	xylene, hexamethylenetetramine,
Resins compounding	1939–1958	lump resins (formaldehyde), accelera-
Molding powder	1939–1956	tors, dry powders, other
Other and unspecified	1939–Present	
Resimines	1946–Present	Melamine, butanol, formaldehyde, urea, resorcinol, other
Formalin plant	1948–Present	Formaldehyde, methanol, silver catalyst, other
Service Dept.-Waste Disposal	1938–Present	Mixed

insignificant compared to other areas, since they were most likely intermittent and very brief.

It should be recognized that workers were not exposed to formaldehyde alone but rather to a mixture of chemical substances of variable nature and at various times. This is a major problem for epidemiological studies in much of the chemical industry. Exposure to chemicals used in the production of formaldehyde, for example, may have been as severe as exposure to formaldehyde itself. Moreover, many of the workers also had exposure to other chemicals prior to or subsequent to their formaldehyde-related job assignment. There seems to be no way of separating exposures, and this study was carried out with the full realization of these possibly confounding factors.

General Mortality Patterns

Three previous unpublished epidemiological studies have examined general mortality patterns among Springfield employees and ex-employees (3-5). All three studies consistently revealed slightly elevated mortality risks for cancer of the digestive system and genitourinary tract; however, definite conclusions could not be drawn due to the lack of information on work history and occupational exposures. The most recent study, conducted by the University of Pittsburgh, Department of Biostatistics, examined mortality during 1950 to 1976 among a cohort of 2490 male wage earners who were employed a minimum of one year between January 1, 1949 and December 31, 1966 (5).

Study Population

The distribution of the Springfield study population by working status and vital status as of December 31, 1976 is shown in Fig. 1. The 603 deaths observed during 1950 through 1976 among this cohort served as the basis for the present proportional mortality study. Included in this decedent group were deaths among retirees ($n = 301$), active employees ($n = 183$), and persons who terminated employment for reasons other than retirement ($n = 119$). All deaths had been coded by an independent nosologist to the Seventh Revision of the International Classification of Diseases and Causes of Death (6).

Work History Data

With the assistance of corporate and plant personnel, complete work history data were obtained and processed for all but one decedent. This decedent plus ten others, who were found on close inspection of work histories to have had less than one cumulative year of employment, were excluded from further study. Detailed analysis of the work history data for the remaining 592 decedents revealed that 136 (23 percent) had one month or more cumulative

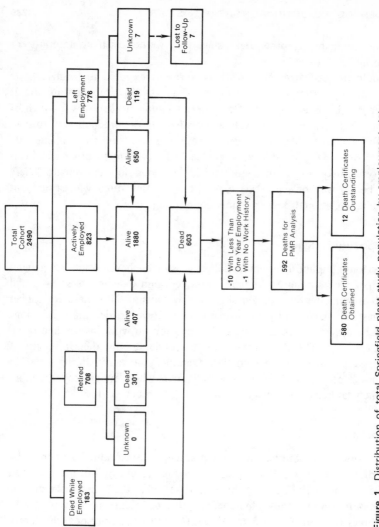

Figure 1 Distribution of total Springfield plant study population by employment status and vital status as of December 31, 1976.

employment in at least one of the five formaldehyde-related plant areas shown in Table 1 and, as such, comprise the "formaldehyde-exposed" group of interest in this report. Death certificates were obtained for 132 (97.1 percent) of the deaths in the formaldehyde-exposed group and for 448 (98.2 percent) of the remaining deaths.

Race

Table 2 shows that 15.4 percent of the formaldehyde-exposed workers are non-white compared to only 2.6 percent among the nonexposed workers. To facilitate the statistical analyses, the six decedents of unknown race were assumed to be white.

RESULTS

Table 3 shows observed deaths and proportional mortality ratios (PMRs) by race for selected causes of death for the formaldehyde-exposed workers and all others, separately. Expected numbers of deaths were calculated by applying the cause-specific proportional mortality of U.S. white and nonwhite males to the total number of white and nonwhite deaths in the study group, while indirectly adjusting for age and time period using the 15 age intervals ($<$ 20, 20-24, 25-29, . . . , 75-79, 80-84, 85+) and six time periods (1950-54, 1955-59, 1960-64, 1965-69, 1970-74, 1975+) (7). PMRs were then calculated by taking the ratio of the sum of the observed number of deaths to the sum of the expected number of deaths ($\times$ 100), summation being taken over age group and time period. The major assumption underlying a proportional mortality analysis is that the study and comparison populations have the same mortality rate for all causes of death. The statistical significance of the difference between observed and expected numbers was assessed by a chi^2 test with one degree of freedom (8). PMRs become unstable with small numbers and were not computed when based on only one observed death.

Table 2 Distribution of Deaths Among Male Chemical Workers by Race and Work Experience

Race	>1 month formaldehyde		All others		Total	
	No.	%	No.	%	No.	%
White	114	83.8	439	96.3	553	93.4
Nonwhite	21	15.4	12	2.6	33	5.6
Unknown	1	0.8	5	1.1	6	1.0
Total	136	100.0	456	100.0	592	100.0

Table 3 Observed Deaths (OBS) and PMRs* During 1950–1976 by Race for Two Groups of Male Chemical Workers Employed a Year or More

Cause of death (I.C.D.A. 7th Revision Codes)	>1 month formaldehyde				All other			
	White		Nonwhite		White		Nonwhite	
	OBS	PMR	OBS	PMR	OBS	PMR	OBS	PMR
All causes (001–999)	115		21		444		12	
All malignant neoplasms (140–205)	20	90.0	2	55.6	96	113.6	5	251.2[†]
Digestive organs and peritoneum (150–159)	8	126.8	1	( )	33	130.2	0	
Esophagus (150)	1	( )	0		3	148.7	0	
Stomach (151)	1	( )	1	( )	9	172.3	0	
Large intestine (153)	1	( )	0		6	79.6	0	
Rectum (154)	1	( )	0		6	197.3	0	
Biliary passages and liver (155–156)	1	( )	0		4	214.1	0	
Pancreas (157)	2	160.2	0		5	104.0	0	
All other digestive organs (residual)	1	( )	0		0		0	
Respiratory system (160–164)	6	79.7	1	( )	20	74.2	1	( )
Bronchus, trachea, lung (162–163)	6	84.7	1	( )	19	75.0	1	( )
Genitourinary tract (177–181)	3	120.5	0		22	192.3[†]	1	( )
Prostate (177)	1	( )	0		12	196.1[†]	1	( )
Bladder (181)	2	320.4	0		3	107.6	0	
Lymphopoietic cancer (200–205)	2	86.3	0		9	108.7	0	
Lymphosarcoma and reticulosarcoma (200)	1	( )	0		1	( )	0	
Leukemia and aleukemia (204)	1	( )	0		2	58.2	0	
All other malignant neoplasms (residual)	1	( )	0		12	75.2	3	882.4[‡]

Cause of death (ICD code)	Obs	PMR	Obs	PMR	Obs	PMR	Obs	PMR
All diseases of nervous system (330–398)	9	110.7	1	( )	27	71.8	2	155.2
All vascular lesions of CNS (330–334)	9	125.9	1	( )	25	72.9	2	170.1
All diseases of circulatory system (400–468)	49	94.4	7	102.2	216	103.5	3	72.6
Arteriosclerotic heart disease (420)	41	96.1	5	111.9	188	110.8	0	
All other circulatory diseases (residual)	8	86.8	2	84.0	28	89.4	3	206.8
All respiratory diseases (470–527)	6	101.1	1	( )	27	108.3	0	
Pneumonia (490–493)	4	178.4	1	( )	14	143.5	0	
Emphysema (527)	1	59.1	0		7	94.1	0	
All other respiratory diseases (residual)	1	( )	0		6	77.5	0	
All diseases of digestive system (530–587)	9	142.4	2	157.5	22	103.8	0	
All external causes of death (800–998)	15	123.0	4	119.1	21	58.1‡	1	( )
Accidents (800–962)	12	148.3	2	106.3	14	56.9†	0	( )
Suicides (963, 970–979)	3	98.6	1	( )	6	68.2	0	
All other causes of death (residual)	4	48.2	3	109.4	27	87.6	1	( )
Unknown causes (in "all causes" category only)	3		1		8		0	

( ) PMRs not shown when based on one observed death.

*Expected numbers of deaths based on U.S. white and nonwhite males.

†p < 0.05.

‡p < 0.01.

243

Table 3 shows that for many of the major cause of death categories examined, the 1950-1976 proportional mortality experience of the formaldehyde-exposed workers was similar to that of the unexposed workers. Generally, PMRs were lower among the formaldehyde-exposed workers for overall cancer mortality and for nonmalignant circulatory and respiratory diseases, whereas PMRs were generally higher among this group for nonmalignant diseases of the nervous and digestive systems and for external causes of death. Curiously, both exposure groups show similarly elevated PMRs for digestive system cancer, whereas the excess genitourinary cancers observed in earlier studies of this plant appear to be concentrated in the nonformaldehyde-exposed group.

For the malignant and nonmalignant diseases examined, no statistically significant excesses or deficits in deaths were observed among white or nonwhite wage earners who worked for a month or more in a formaldehyde-related plant area. For white males an overall 20.3 percent deficit in respiratory cancer deaths was observed. Only one respiratory cancer death was found among the nonwhites. All seven respiratory system cancers observed for the formaldehyde-exposed group were coded to cancer of the bronchus, trachea, or lungs. There have been no deaths from sinonasal cancer among the formaldehyde-exposed or unexposed workers.*

Several statistically significant excesses in deaths were observed among the white and nonwhite nonexposed workers; however, further interpretation of these findings is beyond the scope of this report. Because of the small number of deaths involved among the nonwhites in this study, all subsequent analyses were confined to white males.

Tables 4 and 5 show observed deaths and PMRs for selected causes by age at death and time period, respectively, for the 115 white male formaldehyde-exposed decedents. Of primary interest here are the trends or patterns in PMRs rather than their absolute values. No unusual patterns or trends in proportional mortality by age are discernible in Table 4; however, significantly ($p < 0.05$) elevated PMRs were observed in the youngest age group for digestive cancer (two deaths, PMR = 412.8) and for external causes of death (nine deaths, PMR = 173.6).

Table 5 shows PMRs for digestive system cancer increasing with time, becoming statistically significant ($p < 0.05$) during the time period 1970-1976 (six deaths, PMR = 223.9). A rather pronounced downward trend in PMRs was observed for respiratory system cancer; however, the small numbers of deaths involved does not permit a reliable interpretation of this trend. Table 5 shows a significantly ($p < 0.05$) elevated PMR in the earliest time period examined for "all vascular lesions of the CNS" (three deaths, PMR = 284.6), although this excess does not appear to be associated with any type of trend over time.

---

*In addition, a careful review of all 580 available death certificates revealed no mention whatsoever of sinonasal cancer as a contributory cause of death.

Table 4 Observed Number (OBS) of Deaths and PMRs* by Age at Death Among White Male Formaldehyde-Exposed Chemical Workers

Cause of death (I.C.D.A. 7th Revision Codes)	Age at death					
	< 45		45–64		65+	
	OBS	PMR	OBS	PMR	OBS	PMR
All causes (001–999)	16		64		35	
All malignant neoplasms (140–205)	3	131.0	10	77.0	7	100.8
Digestive organs and peritoneum (150–159)	2	412.8†	3	81.5	3	140.1
Respiratory system (160–164)	0		5	106.5	1	( )
Genitourinary tract (177–181)	0		2	180.1	1	( )
Lymphopoietic tissue (200–205)	1	( )	0		1	( )
Other malignant neoplasms (residual)	0		0		1	( )
All vascular lesions of CNS (330–334)	0		4	122.3	5	147.4
All diseases of circulatory system (400–468)	2	43.3	35	116.9	12	69.3
All respiratory diseases (470–527)	0		2	69.3	4	155.6
All external causes of death (800–998)	9	173.6†	5	85.1	1	( )
All other causes (residual)	2	68.3	6	66.4	5	137.8
Unknown causes (in "all causes" category only)	0		2		1	

( ) PMRs not shown when based on one observed death.

*Expected numbers of deaths based on U.S. white males.

†$p < 0.05$.

Table 5 Observed Number (OBS) of Deaths and PMRs* by Time Period Among White Male Formaldehyde-Exposed Chemical Workers

Cause of death (I.C.D.A. 7th Revision codes)	Time period					
	1950–59		1960–69		1970–76	
	OBS	PMR	OBS	PMR	OBS	PMR
All causes (001–999)	18		47		50	
All malignant neoplasms (140–205)	3	97.0	6	69.7	11	104.4
Digestive organs and peritoneum (150–159)	0		2	77.5	6	223.9†
Respiratory system (160–164)	2	237.3	2	72.2	2	51.1
Genitourinary tract (177–181)	1	( )	1	( )	1	( )
Lymphopoietic tissue (200–205)	0		1		1	
Other malignant neoplasms (residual)	0		0	( )	1	( )
All vascular lesions of CNS (330–334)	3	284.6†	2	65.8	4	130.3
All diseases of circulatory system (400–468)	5	62.9	22	101.2	22	99.1
All respiratory diseases (470–527)	1	( )	3	126.3	2	67.9
All external causes of death (800–998)	4	166.6	7	142.6	4	81.9
All other causes (residual)	2	68.9	5	79.0	6	94.2
Unknown causes (in "all causes" category only)	0		2		1	

( ) PMRs not shown when based on one observed death.
* Expected numbers of deaths based on U.S. white males.
† p < 0.05.

In calculating PMRs for Tables 3, 4, and 5, the expected numbers of deaths were derived from the mortality experience of the entire white or nonwhite male population of the United States. The population in the Springfield, Massachusetts area has a mortality experience that differs somewhat from that of the entire United States, and some of the excesses and deficits in deaths shown in Tables 3, 4, and 5 are due to this difference. Table 6 shows how PMRs differ if based on U.S. proportional mortalities and proportional mortalities computed for the county that surrounds the Springfield area and from which the workforce was largely drawn (Hampden County). It was necessary to restrict the observation period to 1960–1976 in order to make this comparison since proportional mortalities for nonmalignant causes of death were not available for the time period 1950–1959. For most causes of death examined, geographic variability in proportional mortality between the United States and Hampden County appears to be very slight. Notable differences in proportional mortality patterns are evident, however, for digestive system cancer, lymphopoietic cancer, and for external causes of death. Although PMRs based on local mortality patterns may be more valid than those based on the entire United States, they were not used exclusively throughout this report because of the unavailability of a PMR base (all cause mortality) for a significant portion of the study period. In lieu of this deficiency, consideration should be given to the noted geographic

Table 6   Observed Numbers (OBS) of Deaths During 1960–1976[*] Among White Male Formaldehyde-Exposed Chemical Workers, Showing PMRs Based on Both U.S. and Hampden County White Males

		PMR	
Cause of death (I.C.D.A. 7th Revision Codes)	OBS	U.S.	Hampden County
All causes (001–999)	97	100.0	100.0
All malignant neoplasms (140–205)	17	88.8	86.5
Digestive organs and peritoneum (150–159)	8	152.1	130.4
Respiratory system (160–164)	4	59.8	63.9
Genitourinary tract (177–181)	2	91.3	89.7
Lymphopoietic cancer (200–205)	2	102.5	126.8
All other cancer (residual)	1	( )	( )
All vascular lesions of CNS (330–334)	6	98.2	91.6
Arteriosclerotic heart disease (420)	36	98.4	99.0
All respiratory diseases (470–527)	5	94.0	93.6
All external causes of death (800–998)	11	112.3	163.6
All other causes (residual)	19	94.6	84.9
Unknown causes (in "all causes" category only)	3		

( ) PMRs not shown when based on one observed death.

[*]Hampden County proportional mortalities for nonmalignant causes of death were not available for 1950–1959.

variability when interpreting proportional mortality patterns based on the entire United States.

To evaluate the possibility that the overall relative deficit of cancer among white male formaldehyde-exposed workers simply reflects an excess in mortality from other major causes, proportional cancer mortality ratios (PCMRs) were computed for selected cancer sites using age, time, and site-specific proportional cancer mortality to compute expected numbers of deaths. This method assumes that the mortality rate for all cancers was equal in the study and comparison populations and compares the relative frequencies of cancers of specific sites.

Table 7 shows for selected cancer sites the comparison of PMRs and PCMRs based on the U.S. white male mortality experience. With this more conservative approach, proportional cancer mortality during 1950–1976 is somewhat increased; however, the overall excesses and deficits in cancer deaths shown in Table 3 remain statistically nonsignificant ($p > 0.05$).

In order to examine mortality patterns associated with employment in a particular formaldehyde-related plant area, it was necessary to devise a classification scheme that would assign each decedent to the one plant area that best represented his overall formaldehyde exposure experience. To achieve this goal, three different classification schemes were investigated. Scheme 1 categorized the workers into the plant area where first employed. In scheme 2, workers were assigned to the plant area associated with the longest duration of employment. Categorization under scheme 3 was made if a worker was employed in one and only one formaldehyde-related area throughout his entire work history. This third scheme was an attempt to develop "pure" exposure groups, that is, area-specific exposure histories uncontaminated by possibly different formaldehyde exposures in other plant areas.

Table 7   Comparison of Proportional Cancer Mortality Ratios (PCMRs) with Proportional Mortality Ratios (PMRs) During 1950–1976 for White Male Formaldehyde-Exposed Chemical Workers for Selected Cancer Sites

Cause of death (I.C.D.A. 7th Revision Codes)	OBS	PCMR[*]	PMR[†]
All malignant neoplasms (140–205)	20	100.0	90.0
Digestive organs and peritoneum (150–159)	8	141.9	126.8
Respiratory system (160–164)	6	88.4	79.7
Genitourinary tract (177–181)	3	130.4	120.5
Lymphopoietic cancer (200–205)	2	97.6	86.3
All other cancer (residual)	1	( )	( )

( ) PMRs and PCMRs not shown when based on one observed death.

OBS = Observed.

[*]Expected numbers of deaths based on U.S. white male proportional cancer mortality.

[†]Expected numbers of deaths based on U.S. white male proportional mortality (all causes.)

Table 8 Distribution of Deaths Among White Male Formaldehyde-Exposed Chemical Workers by Plant Area According to Three Classification Schemes

Plant area	First job No.	%	Longest duration No.	%	Pure exposure[*] No.	%
Cast phenolics	2	1.7	2	1.7	2	2.4
Resinox	101	87.9	97	84.4	69	83.2
Resins	3	(2.9)	12	(12.4)	2	(2.9)
Resins compounding	0	(0.0)	0	(0.0)	0	(0.0)
Molding powder	93	(92.1)	78	(80.4)	62	(89.9)
Unspecified and other	5	(5.0)	7	(7.2)	5	(7.2)
Resimines	9	7.8	13	11.3	9	10.8
Formalin	2	1.7	2	1.7	2	2.4
Service Dept.-Waste Disposal	1	0.9	1	0.9	1	1.2
Total	115	100.0	115	100.0	83	100.0

[*]Thirty-two (32) employees did not qualify for any pure exposure group.

Table 8 shows the distribution of deaths among the formaldehyde-exposed workers by plant area according to these three classification schemes. Interestingly, all three schemes produced very similar distributions of deaths by plant area. This is probably a reflection of the generally stable nature of the Springfield plant workforce, which was noted in an earlier report (5). Table 8 shows that regardless of the classification scheme employed, over 83 percent of the decedents were categorized into the Resinox area and of these deaths over 80 percent were classified into the subarea of molding powder. None of the schemes categorized any deaths into the Resinox subarea of resin compounding. Unfortunately, the very small numbers of deaths involved does not permit mortality patterns to be reliably examined for individual plant areas other than Resinox. It was of interest, however, to compare the mortality experience of workers employed in plant areas where formaldehyde is present in a liquid form, with potentially higher airborne exposures, to the mortality experience of workers exposed to formaldehyde subsequent to its reaction with other chemicals. Plant area classification scheme 2 (longest duration) was chosen for this comparison primarily because of the better representation of deaths in plant areas other than Resinox.

Table 9 shows observed deaths and PMRs for selected causes for the two groups of workers defined by the nature of their formaldehyde exposure. No statistically significant excesses or deficits in deaths were observed in either exposure group for the causes of death examined. Both groups show similar deficits in total cancer deaths; however, digestive system cancer mortality is more elevated among the workers whose longest exposure was to formaldehyde in the solid form.

Table 9   Observed Numbers (OBS) of Deaths and PMRs[*] During 1950–1976 Among White Male Formaldehyde-Exposed Chemical Workers by Nature of Formaldehyde Exposure

| | Nature of formaldehyde exposure of longest duration | | | |
| Cause of death (I.C.D.A. 7th Revision Codes) | Liquid form[†] | | Solid form[‡] | |
	OBS	PMR	OBS	PMR
All causes (001–999)	36	100.0	79	100.0
All malignant neoplasms (140–205)	6	84.1	14	92.8
Digestive organs and peritoneum (150–159)	2	104.6	6	136.5
Respiratory system (160–164)	1	( )	5	100.2
Genitourinary tract (177–181)	1	( )	2	113.6
Lymphopoietic cancer (200–205)	1	( )	1	( )
Other malignant neoplasms (residual)	1	( )	0	
Vascular lesions of CNS (330–334)	2	98.9	7	136.2
All diseases of circulatory system (400–468)	13	82.0	36	99.9
All respiratory diseases (470–527)	2	109.2	4	97.5
All external causes (800–998)	6	141.5	9	113.2
All other causes (residual)	7	142.6	6	56.1
Unknown causes (in "all causes" category only)	0		3	

( ) PMRs not shown when based on one observed death.

[*]Expected numbers of deaths based on U.S. white males.

[†]Comprised of plant areas: cast phenolics, resins, other Resinox, Resimines, Formalin.

[‡]Comprised of plant areas: molding powder, service dept.

To investigate the possibility of a dose-response relationship between cancer mortality and formaldehyde exposure, PMRs were calculated for the white male formaldehyde-exposed group relative to duration of employment and the interval from the onset of employment until death (latency period). Both duration and latency period were measured with respect to only those jobs held in plant areas associated with formaldehyde exposure. This analysis was done assuming that if cancer deaths were caused by something in the work environment, such as formaldehyde, it would be expected that the greatest excesses would appear among those with the greatest exposure and after some period of time had elapsed.

Table 10 shows that the formaldehyde exposures among the study group were relatively brief since one half were employed less than 2.5 years in one or more of the five formaldehyde-related plant areas. Moreover, these jobs on the average comprised only about one fourth of their overall employment history at the plant. There does appear to have been a sufficient latency period for the development of chronic disease, however, since Table 10 shows that among the 115 decedents an average of 18.8 years elapsed between the initial formaldehyde exposure and death.

Table 10 Characteristics of Employment Among 115 White Male Chemical Worker Deaths (1950-1976)

	Duration of employment (yr)		Latency period (yr)	
	Formaldehyde	Overall	Formaldehyde	Overall
Mean	4.5	16.9	18.8	22.5
Median	2.5	16.3	19.6	20.7
Minimum	0.1	1.2	2.4	2.4
Maximum	18.8	37.0	37.9	53.3

Tables 11, 12, and 13 show observed numbers of deaths and PMRs for selected cancer categories cross-tabulated by the latency period and duration of employment associated with formaldehyde-related jobs. There is no suggestion in the marginal totals of Table 11 that a dose-response relationship exists between formaldehyde exposure and overall cancer mortality among the chemical workers studied. Moderately elevated PMRs were observed within the internal cells of Table 11; however, their pattern relative to latency and duration suggests that these excesses may be due to factors other than formaldehyde exposure.

The marginal totals of Table 12 reveal an unusual pattern of PMRs for digestive system cancer. While over twice as large following 20 years from first employment (compared to less than 20 years), PMRs are only about one half as large among workers with over 5 years of exposure (compared to less than 5 years). If a dose-response relationship existed here, it would be expected that the PMR would be at least as large if not considerably larger for those workers with the greatest duration of exposure. The statistically significant ($p < 0.01$) cluster

Table 11 Observed Deaths (OBS) and PMRs[*] for All Malignant Neoplasms among White Male Formaldehyde-Exposed Chemical Workers During 1950-1976 by Latency Period and Duration of Employment (Measured from First Formaldehyde Job)

Latency (yr)		Duration (yr)		
		<5	5+	Total
<20	OBS	4	5	9
	PMR	47.0	143.8	75.0
20+	OBS	9	2	11
	PMR	155.9	44.8	107.5
Total	OBS	13	7	20
	PMR	91.0	88.2	90.0

[*]Expected number of deaths based on U.S. white males.

Table 12   Observed Deaths (OBS) and PMRs[*] for Digestive System Cancer among White Male Formaldehyde-Exposed Chemical Workers During 1950–1976 by Latency Period and Duration of Employment (Measured from First Formaldehyde Exposure)

Latency (yr)		Duration (yr)		Total
		<5	5+	
<20	OBS	1	2	3
	PMR	41.5	183.3	85.7
20+	OBS	5	0	5
	PMR	320.1[†]	–	178.1
Total	OBS	6	2	8
	PMR	151.1	85.6	126.8

[*]Expected number of deaths based on U.S. white males.
[†]$p < 0.01$.

of digestive system cancer deaths (five observed, PMR = 320.1) within Table 12 may be important or may simply be a chance occurrence arising from the multiple comparisons that were made.[*] It should also be recognized that at least part of the excesses shown in Table 12 are due to the generally higher local proportional mortality for digestive system cancer, which was demonstrated in Table 6.

[*]Four of the five digestive system cancers were categorized Resinox/molding powder according to scheme 2 (longest duration) and one was classified into Resinox/resins.

Table 13   Observed Deaths (OBS) and PMRs[*] for Respiratory System Cancer among White Male Formaldehyde-Exposed Chemical Workers During 1950–1976 by Latency Period and Duration of Employment (Measured from First Formaldehyde Exposure)

Latency (yr)		Duration (yr)		Total
		<5	5+	
<20	OBS	1	2	3
	PMR	35.7	183.8	77.1
20+	OBS	1	2	3
	PMR	47.5	130.4	82.5
Total	OBS	2	4	6
	PMR	40.8	152.6	79.7

[*]Expected number of deaths based on U.S. white males.

   Table 13 shows that respiratory system cancer mortality was substantially higher among workers with 5 or more years of formaldehyde-related employment compared to those employed less than 5 years. In contrast to this finding, however, Table 13 shows an approximate 20 percent deficit in respiratory cancer mortality both before and after 20 years from the onset of the first formaldehyde exposure. Although the very small numbers of deaths involved here do not permit a definitive interpretation to be made, these two findings are inconsistent and do not support the existence of a dose-response relationship between formaldehyde exposure and respiratory cancer.

DISCUSSION

The present study has certain limitations that make it difficult to draw definite conclusions. First, risk estimates obtained from PMR analyses only approximate the results from studies of cause-specific disease rates. Although PMRs are often similar to risk estimates when the population at risk is available (9), they may be inflated for certain causes when the overall mortality of the study group is lower than that of the comparison population, as is usually the case with working populations (10).

   Second, a special problem for studies of chemical plant workers is uncertainty as to just what exposures and combination of exposures took place. It is likely that most of the workers studied in this report were not exposed to formaldehyde alone but rather to a great number of chemicals that have been used in the manufacture of the major products made at the Springfield plant. A partial list of other chemicals used or produced during the history of operations includes phenol, cellulose acetate, polyvinyl butyral, amino plastics, polystyrene, and various vinyl chloride products, including vinyl chloride polymers. Although established chronic disease risks have not been associated with most of these chemicals, exposure to vinyl chloride and its polymers has been linked with elevated risks for angiosarcomas of the liver, digestive and respiratory system cancers, brain cancers, and lymphomas (11-13). Vinyl chloride exposures most likely did not occur in the plant areas associated with formaldehyde; however, it is possible that some men were exposed to this chemical at some time during their work history at the plant. It is uncertain at this time whether these possible vinyl chloride exposures in any way affected the proportional mortality patterns among the workers studied in this report.

   Third, examining the proportional mortality experience of the formaldehyde-exposed workers relative to race, calendar time, age, duration of employment, and latency period generally resulted in very small numbers of deaths within most subcategories. This was particularly true for malignant neoplasms, which were studied in great detail. The small numbers of deaths involved in these subcategories often made trends or patterns in mortality difficult to discern. This was a particular hindrance in this study, for without knowledge of actual historical formaldehyde exposure levels, these trends and patterns provided the only means of systematically evaluating mortality in relation to occupational factors.

Small sample size was also a problem in examining proportional mortality relative to the nature of the formaldehyde exposure. Over 68 percent of the white male study group spent the largest portion of their formaldehyde-related employment in plant areas where formaldehyde existed in combination with other chemicals as a solid or powdered substance. Except for rare instances, airborne exposures to formaldehyde vapor in these areas were probably very low. This means that only 36 members of this study group worked for the majority of their employment history in areas where significant formaldehyde vapor exposures could have occurred on a routine basis. Clearly, further studies of much larger groups of such workers are needed before an accurate and reliable assessment can be made of the possible mortality risks associated with formaldehyde exposures.

Despite the aforementioned limitations, the results of this study did not reveal any trends or patterns in proportional mortality that could be directly linked to occupational exposures to formaldehyde at the Springfield plant.

SUMMARY

A proportional mortality analysis was conducted on deaths occurring between 1950 and 1976 among 136 males who were employed a month or more in one of five formaldehyde-related areas of a large chemical-producing plant. Overall, no statistically significant excesses or deficits in proportional mortality were observed among the formaldehyde-exposed group based on comparisons with both U.S. males and males from the local county area. In addition, no important differences in mortality were observed among this group when comparisons were made with 456 male decedents from the same plant who did not have a month or more of formaldehyde exposure. No sinonasal cancer deaths were observed among the chemical workers studied nor was mention made on any death certificate of sinonasal cancer as a contributory cause of death. No important excesses, trends, or patterns in cancer mortality were observed among white male formaldehyde-exposed workers when consideration was given to age and time period of death, type and duration of exposure, and the lapsed period from the onset of the first formaldehyde-related job assignment. Although certain limitations of this study do not allow definite conclusions to be drawn, the results indicate no trends or patterns in proportional mortality that could be directly linked to formaldehyde exposures.

REFERENCES

1 *Criteria for a Recommended Standard . . . Occupational Exposure to Formaldehyde.* HEW (NIOSH) Publication No. 77-126 (1976).

2 Proctor, NH, and JP Hughes. *Chemical Hazards of the Workplace*, Philadelphia: Lippincott (1978).

3 Stanislawczyk, K, R Kaminski, and R Spirtas. A proportional mortality analysis of a chemical plant in Massachusetts. NIOSH report (May 1978).

4 Proportional mortality study of Monsanto-Indian Orchard plant employees. Monsanto report (1978).

5 Marsh GM. Final report on the Monsanto Company Indian Orchard plant mortality study. University of Pittsburgh, Department of Biostatistics Technical Report (December 1979).

6 National Center for Health Statistics. *Seventh Revision International Classification of Diseases Adapted for Use in the United States*, Vol. 1, Tabular List; Vol. 2, Alphabetical Index (1957).

7 Marsh, GM, and M Preininger. OCMAP: A user-oriented occupational cohort mortality analysis program. *The American Statistician* 34:210–212 (1980).

8 Miettinen, OS. Estimability and estimation in case-referent studies. *Am. J. Epidemiol.* 103:226–235 (1976).

9 Kupper, LL, AJ McMichael, MJ Symons, and BM Most. On the utility of proportional mortality analysis. *J. Chronic Dis.* 31:15–22 (1978).

10 McMichael, AJ. Standardized mortality ratios and the "healthy worker effect": Scratching beneath the surface. *J. Occup. Med.* 18:165–168 (1976).

11 Tabershaw, I, and W Gaffey. Mortality study of workers in the manufacture of vinyl chloride and its polymers. *J. Occup. Med.* 16:509–518 (1974).

12 Waxweiler, R, W Stringer, J Wagoner, et al. Neoplastic risk among workers exposed to vinyl chloride. *Ann. NY Acad. Sci.* 271:40–48 (1976).

13 Monson, R, and J Peters. Proportional mortality among vinyl chloride workers. *Lancet* 2:397–398 (1974).

Chapter 22

An Epidemiologic Mortality Study of a Cohort of Chemical Workers Potentially Exposed to Formaldehyde, with a Discussion on SMR and PMR

Otto Wong

Formaldehyde is perhaps one of the most important chemicals in the production of thousands of industrial and commercial products. Some of the major industries that use formaldehyde are plastics, resins, adhesives, sealants, textiles, plywood, particleboard, wood, paper, foundries, funeral homes, paints, dyes, and pigments (1).

Formaldehyde has long been known to be an eye and respiratory irritant and has also been demonstrated as a contact allergen. In a recent Chemical Industry Institute of Toxicology (CIIT) formaldehyde inhalation study, rats and mice were exposed to various concentrations of formaldehyde. After 18 months of exposure, a high incidence of squamous cell carcinomas in the nasal turbinates was found among rats exposed to 15 ppm formaldehyde (2). At this point, however, there is no epidemiologic evidence that formaldehyde is a human carcinogen. Obviously, such epidemiologic data are badly needed.

MATERIALS AND METHODS

Data for this investigation came from a chemical plant that is one of the largest producers of formaldehyde in this country. The plant was built in the early

1940s, and formaldehyde has been manufactured at the plant since that time. However, in addition to formaldehyde, a variety of products and processes have been introduced over the years. Potential exposures at the plant (some no longer exist) include formaldehyde and other oxygenated hydrocarbons, benzene, asbestos, and inorganic and organic pigments.

The design of the epidemiologic investigation is that of an historical prospective mortality study. The study includes all white male workers ever employed at the plant between start-up and 1977. Data on females employed at the plant during the same time period have also been collected. The number of females is relatively small (slightly over 200), and since these female employees would not have been directly exposed to the chemicals, they will not be included in the present paper. In addition, 12 blacks and 2 Orientals were also excluded from the analysis because of the small number and the considerable difference in mortality between the nonwhites and the whites. However, it should be pointed out that no death was observed among these nonwhites. Included in the analysis were 2026 white males employed at the plant through 1977.

Data collected on each cohort member included name, social security number, date of birth, race, sex, date of hire, date of separation, employment status, vital status (as of December 31, 1977), date and place of death (if applicable), and work history. Each work history consisted of a chronological listing of all jobs ever held by the individual. Information included date, department, plant area, unit description, job classification, job title, and a complete list of potential contaminants associated with the job. The coding of work history is still under way, and will be completed in a month or two. As such, this analysis will not consider detailed analysis by work history or exposure.

Among the 2026 workers in the cohort, 791 were active on December 31, 1977. It was necessary to determine whether the 1235 workers who had terminated their employment prior to December 31, 1977 were still living or whether they had died in the interim between the termination of their employment and the end of the study. In order to expedite the study, names and social security numbers of all terminated employees were key-punched prior to any coding to generate a list of individuals who needed vital status follow-up. This list was sent to the Social Security Administration (SSA) for a search through the SSA files. The SSA follow-up provided information as to whether the individual was still living (paying contributions or receiving benefits), had died (a death benefit claim has been filed), or unknown to SSA. The names and birthdates of those individuals whose vital status was unknown to SSA were sent to the state motor vehicle bureau. Persons who renewed their driving licenses after December 31, 1977 were considered living. Those who renewed their driving licenses prior to December 31, 1977 were contacted by telephone whenever possible to ascertain their vital status on December 31, 1977. Finally, a local follow-up by plant personnel was also carried out. The local vital status follow-up consisted of having several long-term employees familiar with the workers and the community review the list of the remaining unknowns. For those identified as living by the local follow-up, it was required to record their addresses or telephone numbers so that the information could be verified.

Table 1 provides a summary of the vital status of the entire cohort. Of the 2026 male workers in the cohort, 1829 (90.28 percent) were still living as of December 31, 1977; 146 (7.21 percent) were identified as deceased; and 51 (2.52 percent) were "lost to follow-up" (vital status unknown). For the 146 decedents, date of death and place of death were obtained through the SSA or the local follow-up. Requests for death certificates were made to state vital records departments. Death certificates were obtained for all decedents except for 10 cases (6.85 percent).

Causes of death were coded by a trained nosologist according to the 8th revision of the *International Classification of Diseases* (3). A total of 136 deaths among the white males were coded.

The most common summary index for assessing the risk of death in a population studied prospectively is the standardized mortality ratio (SMR). Basically, the number of deaths occurring in the study population is compared with the number of deaths that would have been expected if the study population had had, adjusting for age and time period, the same mortality experience as a comparable nonexposed population. During the observation period of the study, the cohort members entered the study at different points in time; some died; and some were lost to follow-up. For all these reasons, cohort members are observed for various lengths of time and, therefore, do not contribute equally to the "population at risk." To utilize fully the period of observation for each member and to properly weight the SMR, the concept of "person-years" is used in the analysis. The basic unit of computation is the number of years each employee was followed, from the date of first employment to the end of the study period or to the date of death. Each year contributed by an individual worker is classified by age and calendar year, and these person-years are then

Table 1 Vital Status (as of December 31, 1977) of 2026 White Male Chemical Workers

Vital status	Frequency	%
Living	1829	90.28
(Active)	(791)	(43.25)
(Terminated)	(1038)	(56.75)
Deceased	146	7.21
(With death certificates)	(136)	(93.15)
(Without death certificates)	(10)	(6.85)
Unknown	51	2.52
Total	2026	

Table 2   Person-Years of Observation by Age and Calendar Year of 2026 White Male Chemical Workers

Age	<1950	1950-	1955-	1960-	1965-	1970-	1975-	Sum
<20	16.5	30.4	20.4	15.1	13.8	15.4	10.5	121.8
20-	176.5	413.7	247.2	138.7	262.4	252.1	299.3	1790.0
25-	267.1	905.7	849.3	538.5	665.7	940.5	500.3	4667.1
30-	188.3	737.5	1132.1	988.8	805.7	966.6	741.6	5560.6
35-	171.8	461.3	900.5	1190.3	1101.0	927.7	592.9	5345.5
40-	111.1	307.8	520.6	956.2	1244.6	1200.2	590.8	4931.3
45-	74.1	179.9	313.7	522.5	960.1	1264.6	772.9	4087.7
50-	37.2	109.3	190.0	294.3	512.7	949.3	727.9	2820.6
55-	11.8	50.1	110.0	183.1	287.4	496.5	486.2	1625.2
60-	5.8	18.4	48.6	103.4	167.9	271.0	250.8	866.0
65-	0.0	7.8	18.4	44.6	90.2	145.6	121.1	427.7
70-	0.0	0.0	6.8	15.7	34.2	69.2	64.2	190.1
75-	0.0	0.0	0.0	5.3	8.9	22.3	29.6	66.0
80-	0.0	0.0	0.0	0.0	5.3	2.0	5.5	12.7
85+	0.0	0.0	0.0	0.0	0.0	2.2	0.0	2.2
Sum	1059.4	3221.9	4357.5	4996.4	6159.7	7525.1	5193.6	32514.3

summed up by age and calendar year (Table 2). The U.S. national age, sex, year, cause-specific mortality rates for 5-year time periods from 1925–1975 are applied to these person-years to obtain the number of deaths from a particular cause to be expected from an equal number of person-years similar in age, sex, and calendar year. SMRs can be computed by expressing the actually observed deaths as percentages of the expected.

An SMR higher (lower) than 100 indicates an excess (a deficit) in mortality. The deviation from 100 is tested to determine whether it is statistically significant. The statistical analysis was performed using a standard computer program (4). This program computes an SMR for every cause of death included, even if only one death is attributed to that cause. Some epidemiologists and biostatisticians prefer not to calculate SMR when the number of deaths is two or less (or even five or less) because of the large amount of variability associated with such a small number of deaths. The approach in this report is to provide all SMRs and their confidence intervals. If the number of deaths is small, the corresponding variance, and hence the confidence interval, will be large. In this case, unless the SMR is extremely high, the result will not be statistically significant. As such, the significance test takes into account the small number involved. However, one concern is that for rare diseases, 1 or 2 cases will be adequate for statistical significance. While for rare diseases expected deaths can be calculated down to a small fraction, in reality, deaths occur in whole numbers. In this sense, statistically significant SMRs with only one or two observed deaths should be interpreted with caution.

RESULTS

Mortality of Entire Cohort

The observed deaths, expected deaths, and SMRs by cause for the entire cohort are listed in Table 3. The limits of the 95 percent confidence intervals for the SMRs are also presented. If the lower limit is higher than 100, there is a significant mortality excess at the 0.05 level. If the upper limit is lower than 100, there is a significant deficit. Similarly the 99 percent confidence intervals have also been calculated for selected SMRs but are not presented in Table 3. A total of 146 deaths were included in the analysis. As mentioned earlier, death certificates were not obtained for 10 cases. These 10 deaths were included in the overall SMR but not in any cause-specific SMRs. Since 196.85 deaths were expected, the overall SMR is 74, indicating that these chemical workers enjoyed a mortality deficit of 26 percent, compared to the U.S. white males. This deficit is statistically significant at the 0.01 level, since the upper limit of the 99 percent confidence interval is less than 100. The mortality deficit appears to have come from diseases of the circulatory system. A total of 53 deaths in the cohort were coded as diseases of the circulatory system, while 87.97 deaths were expected based on the U.S. white male experience, adjusted for age and calendar time. The corresponding SMR for diseases of the circulatory system is 60, statistically significant at the 0.01 level. Within the broad category of diseases of the circulatory system, mortality from arteriosclerotic heart disease among these chemical workers was significantly less than the U.S. white males (35 observed versus 64.26 expected, SMR = 54, $p < 0.01$).

In addition to diseases of the circulatory system, the mortality deficit in nonmalignant respiratory diseases was also statistically significant. Three deaths were ascribed to respiratory diseases, compared to an expectation of 9.37. The corresponding SMR is 32, $p < 0.05$. Similarly, mortality from diseases of the digestive system among these chemical workers was significantly less than the U.S. white males. SMR for diseases of the digestive system is 26 (3 observed versus 11.48 expected, $p < 0.05$).

Mortality from cancer of all sites for the entire cohort was identical to the expectation (37 observed versus 36.50 expected, SMR = 101, nonsignificant). None of the site-specific cancer SMRs is significant. In particular, mortality from cancer of the respiratory system was exactly the same as expected (SMR = 97). No nasal cancer death was observed. There was a mortality deficit in cancer of the digestive system. However, this deficit is not statistically significant. One death was due to cancer of the bone, more than four times the expected; but the excess is not statistically significant. It should also be pointed out that four deaths were attributed to cancer of the prostate, when 1.31 was expected. Although the risk ratio is threefold (SMR = 305), the excess is not significant at the 0.05 level. Nonsignificant mortality excesses were also observed in cancer

Table 3 Observed and Expected Deaths by Cause, SMRs, and Their 95 Percent Confidence Limits for the Entire Cohort of 2026 White Male Chemical Workers[*]

Cause of death (8th Revision I.C.D.A.)	Observed deaths	Expected deaths	SMR	Lower limit	Upper limit
All causes	146	196.85	74[‡]	63	87
All cancers (140–209)	37	36.50	101	71	140
Cancer of digestive system (150–159)	5	9.46	53	17	123
Cancer of stomach (151)	1	1.77	56	1	313
Cancer of large intestine (153)	3	2.96	101	20	296
Cancer of pancreas (157)	1	1.92	52	1	289
Cancer of respiratory system (160–163)	12	12.36	97	50	170
Cancer of lung (162–163)	11	11.67	94	47	169
Cancer of bone (170)	1	0.23	430	6	2393
Cancer of skin (172)	1	0.92	109	1	607
Cancer of prostate (185)	4	1.31	305	82	780
Cancer of bladder (188)	1	0.82	122	2	681
Cancer of kidney (189)	1	0.98	102	1	570
Cancer of brain (191)	3	1.61	186	37	545
Lymphopoietic cancer (200–209)	6	4.42	136	50	295
Hodgkin's disease (201)	2	0.83	240	27	866
Leukemia and aleukemia (204–207)	2	1.70	118	13	426
Diseases of circulatory system (390–458)	53	87.97	60[‡]	45	79
Arteriosclerotic heart disease (inc. CHD) (410–413)	35	64.26	54[‡]	38	76
Cerebrovascular disease (430–438)	13	9.67	134	71	230
Nonmalignant respiratory disease (460–519)	3	9.37	32[†]	6	94
Emphysema (492)	2	2.30	87	10	314
Diseases of digestive system (520–577)	3	11.48	26[†]	5	76
Cirrhosis of liver (571)	3	6.73	45	9	130
Symptoms, senility, and ill-defined conditions (780–796)	2	2.56	78	9	283
Accidents, poisonings, and violence (E800–E999)	33	34.52	96	66	134
Accidents (800–949)	26	23.60	110	72	161
Motor vehicle accidents (810–827)	18	12.07	149	88	236
Suicide (950–959)	5	7.56	66	21	154

[*]Person-years = 32,514.3.
[†]Statistically significant at 0.05 level.
[‡]Statistically significant at 0.01 level.

Table 4   Observed and Expected Deaths by Cause, SMRs, and Their 95 Percent Confidence Limits for 998 White Male Chemical Workers Hired Prior to December 31, 1960[*]

Cause of death (8th Revision I.C.D.A.)	Observed deaths	Expected deaths	SMR	Lower limit	Upper limit
All causes	129	176.99	73[‡]	61	87
All cancers (140–209)	34	33.74	101	70	141
Cancer of digestive system (150–159)	4	8.92	45	12	115
Cancer of stomach (151)	1	1.69	59	1	330
Cancer of large intestine (153)	2	2.77	72	8	261
Cancer of pancreas (157)	1	1.81	55	1	307
Cancer of respiratory system (160–163)	10	11.62	86	41	158
Cancer of lung (162–163)	9	10.96	82	37	156
Cancer of bone (170)	1	0.20	494	6	2751
Cancer of skin (172)	1	0.78	129	2	717
Cancer of prostate (185)	4	1.28	311	84	797
Cancer of bladder (188)	1	0.79	127	2	705
Cancer of kidney (189)	1	0.91	110	1	613
Cancer of brain (191)	3	1.41	213	43	623
Lymphopoietic cancer (200–209)	6	3.87	155	57	338
Hodgkin's disease (201)	2	0.68	294	33	1063
Leukemia and aleukemia (204–207)	2	1.48	135	15	487
Diseases of circulatory system (390–458)	49	82.97	59[‡]	44	78
Arteriosclerotic heart disease (inc. CHD) (410–413)	31	60.68	51[‡]	35	73
Cerebrovascular disease (430–438)	13	9.14	142	76	243
Nonmalignant respiratory disease (460–519)	3	8.77	34	7	100
Emphysema (492)	2	2.22	90	10	326
Diseases of digestive system (520–577)	3	10.36	29[†]	6	85
Cirrhosis of liver (571)	3	6.00	50	10	146
Symptoms, senility, and ill-defined conditions (780–796)	2	2.18	92	10	331
Accidents, poisonings, and violence (E800–E999)	23	26.00	88	56	133
Accidents (800–949)	19	17.99	106	64	165
Motor vehicle accidents (810–827)	11	8.82	125	62	223
Suicide (950–959)	2	5.87	34	4	123

[*]Person-years = 24,890.7.
[†]Statistically significant at 0.05 level.
[‡]Statistically significant at 0.01 level.

of the brain and other parts of the central nervous system and in lymphopoietic cancer. In fact, no significant mortality excess was found in the entire cohort as a whole for any of the causes of death examined.

Mortality by Date of Hire

Tables 4 and 5 present the mortality experience for those hired prior to December 31, 1960 and those hired after January 1, 1961, respectively. The year 1960 is roughly the midpoint of the observation period and divides the cohort into two groups of the same size. Among the group hired prior to December 31, 1960, a total of 129 deaths were observed, compared to 176.99 (SMR = 73, $p < 0.01$). The total mortality, as well as the cause-specific mortality, in this group was very similar to that of the entire cohort. Significant mortality deficits were found for diseases of the circulatory system, arteriosclerotic heart disease, and diseases of the digestive system among those hired prior to December 31, 1960. Mortality from all individual causes of death examined was within the expected range among the same group of workers. No significant cause-specific mortality excess was detected.

Among those workers hired after January 1, 1961, only 17 deaths occurred. The expected number of deaths was 19.86, and the corresponding overall SMR

Table 5 Observed and Expected Deaths by Cause, SMRs, and Their 95 Percent Confidence Limits for 1028 White Male Chemical Workers Hired after January 1, 1961[*]

Cause of death (8th Revision I.C.D.A.)	Observed deaths	Expected deaths	SMR	Lower limit	Upper limit
All causes	17	19.86	86	50	137
All cancers (140–209)	3	2.75	109	22	318
Cancer of digestive system (150–159)	1	0.54	186	2	1033
Cancer of large intestine (153)	1	0.19	529	7	2943
Cancer of respiratory system (160–163)	2	0.75	267	30	966
Cancer of lung (162–163)	2	0.71	282	32	1018
Diseases of circulatory system (390–458)	4	5.00	80	22	205
Arteriosclerotic heart disease (inc. CHD) (410–413)	4	3.57	112	30	287
Accidents, poisonings, and violence (E800–E999)	10	8.52	117	56	216
Accidents (800–949)	7	5.61	125	50	257
Motor vehicle accidents (810–827)	7	3.25	215	86	444
Suicide (950–959)	3	1.70	177	36	517

[*]Person-years = 7623.7.

is 86 (nonsignificant). All of the cause-specific SMRs presented in Table 5 are within the expected range. One reason for this observation may have been the small number of deaths.

Mortality by Latency

For many chronic diseases, there is usually a long latent period between first exposure and onset of disease or death. Therefore, it would be logical to examine mortality experience after a certain lag period has elapsed. Cause-specific SMRs by latency for the entire cohort of chemical workers are presented in Tables 6 and 7. When mortality was analyzed after a latent period of 10 years had elapsed, it was found that total mortality was significantly less than the expectation ($p < 0.05$). Other than total mortality, all the cause-specific SMRs presented in Table 6 are within the expected range.

When a latent period of 20 years is considered, SMR for total mortality becomes 83 (nonsignificant). Mortality from cancer of the respiratory system was slightly, but not significantly, less than the expected (Table 7). For prostatic cancer, the number of observed deaths was 4 and the number of expected deaths was 0.93. The corresponding SMR for prostatic cancer is 431, significant at the 0.05 level. No other site-specific cancer mortality is significant. In the same group, mortality from motor vehicle accidents was significantly higher than the expected (6 observed versus 1.80 expected, SMR = 333, $p < 0.05$). Except for prostatic cancer and motor vehicle accidents, all the cause-specific SMRs are within the expected range, even when a latency of 20 years is taken into account.

Mortality by Length of Employment

Table 8 provides a mortality analysis by length of employment for the entire cohort. Such an analysis will eliminate the "dilution effect" of including short-term employees and will also indicate whether a longer employment at the plant could in fact be associated with a more adverse cause-specific mortality. It should be pointed out that the basic unit of calculation in this analysis is person-year. For example, an employee left after 9 years of service and subsequently died; the first 5 years would be included in the first length of employment group (<5 years), and the remaining years till death would be included in the second length of employment group (5-9 years). In this case, the subsequent death would be counted in the second length of employment group.

As Table 8 clearly demonstrates, there is no indication of any trend between mortality and length of employment for all causes, cancer of the respiratory system, cancer of the prostate, diseases of the circulatory system, and motor vehicle accidents. Among all the cause-specific SMRs in Table 8, only the one for cancer of the prostate for those with less than 5 years of employment is

Table 6   Observed and Expected Deaths by Cause, SMRs, and Their 95 Percent Confidence Limits for 1279 White Male Chemical Workers after a Latency Period of 10 Years[*]

Cause of death (8th Revision I.C.D.A.)	Observed deaths	Expected deaths	SMR	Lower limit	Upper limit
All causes	115	150.48	76[†]	63	92
All cancers (140–209)	30	30.43	99	67	141
Cancer of digestive system (150–159)	3	7.97	38	8	110
Cancer of large intestine (153)	2	2.52	79	9	286
Cancer of pancreas (157)	1	1.68	60	1	332
Cancer of respiratory system (160–163)	11	11.04	100	50	178
Cancer of lung (162–163)	10	10.45	96	46	176
Cancer of bone (170)	1	0.15	662	9	3685
Cancer of prostate (185)	4	1.23	324	87	830
Cancer of bladder (188)	1	0.73	137	2	762
Cancer of kidney (189)	1	0.83	120	2	669
Cancer of brain (191)	2	1.16	173	19	625
Lymphopoietic cancer (200–209)	4	3.17	126	34	323
Hogdkin's disease (201)	1	0.45	223	3	1239
Leukemia and aleukemia (204–207)	2	1.20	167	19	603
Diseases of circulatory system (390–458)	49	74.46	66	49	87
Arteriosclerotic heart disease (inc. CHD) (410–413)	31	55.60	56	38	79
Cerebrovascular disease (430–438)	13	8.14	160	85	273
Nonmalignant respiratory disease (460–519)	3	7.85	38	8	112
Emphysema (492)	2	2.16	93	10	334
Diseases of digestive system (520–577)	3	8.97	33	7	98
Cirrhosis of liver (571)	3	5.44	55	11	161
Symptoms, senility, and ill-defined conditions (780–796)	2	1.95	102	11	369
Accidents, poisonings, and violence (E800–E999)	17	17.39	98	57	156
Accidents (800–949)	13	11.24	116	62	198
Motor vehicle accidents (810–827)	7	5.12	137	55	282
Suicide (950–959)	2	4.43	45	5	163

[*]Person-years = 16,157.8.
[†]Statistically significant at 0.05 level.

significantly higher than the expectation. There seems to be a general decreasing trend in SMRs for cancer of all sites with an increasing length of employment. The trend, however, is not monotonic.

There appears to be a monotonically increasing trend of lymphopoietic cancer mortality with an increasing length of employment, starting with an SMR of 77 in the shortest employment group to an SMR of 210 in the longest em-

Table 7   Observed and Expected Deaths by Cause, SMRs, and Their 95 Percent Confidence Limits for 866 White Male Chemical Workers after a Latency of 20 Years[*]

Cause of death (8th Revision I.C.D.A.)	Observed deaths	Expected deaths	SMR	Lower limit	Upper limit
All causes	72	86.51	83	65	105
All cancers (140–209)	21	18.88	111	69	170
Cancer of digestive system (150–159)	2	4.88	41	5	148
Cancer of large intestine (153)	2	1.62	124	14	447
Cancer of respiratory system (160–163)	6	7.24	83	30	180
Cancer of lung (162–163)	6	6.87	87	32	190
Cancer of prostate (185)	4	0.93	431[†]	116	1103
Cancer of bladder (188)	1	0.49	203	3	1129
Cancer of brain (191)	1	0.59	170	2	945
Lymphopoietic cancer (200–209)	4	1.73	231	62	591
Hodgkin's disease (201)	1	0.17	582	8	3236
Leukemia and aleukemia (204–207)	2	0.65	306	34	1104
Diseases of circulatory system (390–458)	32	44.89	71	49	101
Arteriosclerotic heart disease (inc. CHD) (410–413)	20	33.84	59	36	91
Cerebrovascular disease (430–438)	8	5.10	157	68	309
Nonmalignant respiratory disease (460–519)	1	5.05	20	0	110
Emphysema (492)	1	1.47	68	1	378
Diseases of digestive system (520–577)	3	4.80	63	13	183
Cirrhosis of liver (571)	3	2.95	102	20	297
Symptoms, senility, and ill-defined conditions (780–796)	2	1.11	180	20	649
Accidents, poisonings, and violence (E800–E999)	9	6.88	131	60	248
Accidents (800–949)	9	4.31	209	95	397
Motor vehicle accidents (810–827)	6	1.80	333[†]	122	725

[*]Person-years = 5947.6.
[†]Statistically significant at 0.05 level.

Table 8 Observed Deaths and SMRs by Cause for the Entire Cohort of White Male Chemical Workers by Length of Employment

Cause of death (8th Revision I.C.D.A.)		Length of employment (yr)				
		<5	5-9	10-14	15-19	20+
All causes	Deaths	51	16	17	22	33
	SMR	96	60	58	78	70
All cancers (140-209)	Deaths	14	5	5	6	5
	SMR	168	112	96	109	47
Cancer of respiratory system (160-163)	Deaths	6	1	1	3	0
	SMR	240	74	62	157	0
Cancer of prostate (185)	Deaths	2	0	0	1	1
	SMR	1053*	0	0	381	221
Lymphatic and hematopoietic cancer (200-209)	Deaths	1	1	1	1	2
	SMR	77	153	164	175	210
Diseases of circulatory system (390-458)	Deaths	11	3	10	9	18
	SMR	58	28	68	62	74
Motor vehicle accidents (810-827)	Deaths	10	5	0	0	3
	SMR	172	237	0	0	308

*Statistically significant at 0.05 level.

ployment group. However, it must be emphasized that for four of the five groups, only one death was involved, and only two deaths in the remaining group. As such, the statistical variability associated with each SMR is substantial, and the result must be viewed with caution.

DISCUSSION AND CONCLUSIONS

There are several advantages of using cause-specific mortality as a measure of ill health. Deaths are well defined and have been reliably and consistently recorded, and such records are readily accessible. In general, mortality provides a good measure for incidence of a disease if the condition is fairly fatal. Furthermore, in many instances, only mortality studies are feasible, especially when rare chronic diseases with long latent periods are involved. In occupational mortality studies, the most common indexes are the standardized mortality ratio (SMR) and the proportionate mortality ratio (PMR).

In a prospective (historical or forward) study, a cohort is defined and the number of deaths is observed over a period of time. A mortality rate can be computed by dividing the number of deaths by the number of person-years at risk. The SMR, which is a particular measure of relative risk, is simply the ratio of the mortality rate of an exposed cohort to that of an unexposed control group, adjusted for age, race, and other confounding variables.

However, in many cases, the population at risk cannot be defined because of inadequate records, inaccessible records, or limited resources. If the number of deaths and the causes of death are known, the cause-specific mortality proportion, defined as the percentage of deaths from a given cause to the total deaths, can be computed. The PMR is the ratio of the cause-specific mortality proportion in an exposed group to that in an unexposed group, adjusted for age, race, and other confounding variables. A PMR greater than unity (or 100 percent) indicates that there is a higher proportion of deaths due to a specific cause in the study group than in the controls. This may or may not mean a corresponding higher *risk* for the same cause. It can be shown that the PMR is equivalent to the SMR only if the age-specific mortality rates (all causes) in the study group are the same as those in the controls (see the Appendix). Nevertheless, PMR has been used (misused?) by many investigators for risk assessment. Since most occupational groups have a lower overall total mortality, PMR tends to overestimate the mortality experience.

Table 9 shows selected SMRs and PMRs for the entire cohort. PMR is significant for all cancers, cancer of the bone, cancer of the prostate, cerebrovascular disease, all external causes of death, and all accidents and motor vehicle accidents. Of course, none of the corresponding SMRs is significant, and if PMRs are used for risk assessment, they would give erroneous results.

Recently, a method of converting PMRs to "SMRs" has been proposed (5). The proposed approach assumes that the age-specific overall SMRs remain constant for all age-groups. This assumption is most unrealistic. Based on data from the Registrar General's Occupational Decennial Supplement to the 1951 Census of England and Wales, Kilpatrick divides the entire English and Welsh population

Table 9   Observed Deaths, SMRs, and PMRs for Selected Causes for the Entire Cohort of 2026 White Male Chemical Workers

Cause of death (8th Revision I.C.D.A.)	Observed deaths	SMR	PMR
All cancers (140–209)	37	101	148*
Cancer of respiratory system (160–163)	12	97	143
Cancer of bone (170)	1	430	625*
Cancer of prostate (185)	4	305	367*
Lymphatic and hematopoietic cancer (200–209)	6	136	194
Diseases of circulatory system (390–458)	53	60†	87
Arteriosclerotic heart disease (410–413)	35	54†	79
Cerebrovascular disease (430–438)	13	134	183*
All external causes of death (800–999)	33	96	131*
All accidents (800–949)	26	110	150*
Motor vehicle accidents (810–827)	18	149	194*

*Statistically significant at 0.05 level.
†Statistically significant at 0.01 level.

into 14 broad occupational groups and tests for heterogeneity of age-specific overall SMRs within each group (6). Of the 14 groups, the chi^2 for heterogeneity has been found to be statistically significant at the 0.001 level for 13 groups. As such, from the methodological point of view, PMR is unsuitable for risk assessment.

From the practical point of view, there is yet another limitation of PMR. A practical issue that needs extensive consideration is the usual situation in which occupational PMR studies are most likely carried out. Most often occupational PMR studies are based exclusively on deaths known to either the employers or labor unions. Even within the framework of a PMR analysis, these studies are potentially biased since deaths among terminated employees are not included. In a mortality study of 59,000 steelworkers (7), it was found that the proportion of cancer deaths unknown to the employer (45 percent) was more than twice that of cardiovascular diseases (21 percent). Since by definition, PMR is based on proportional mortality analysis by cause, such a potential bias resulting from a skewed distribution of deaths by cause derived from an incomplete ascertainment of deaths should not be overlooked lightly.

Based on the above discussion, it is clear that SMR, and not PMR, is the index of choice in risk assessment, and PMR in most instances would only confuse the issues. However, PMR can be a useful epidemiologic tool in detecting unusual clusters of cause-specific deaths for surveillance purposes.

Using the SMRs in this report as the criteria, a number of conclusions regarding the mortality of this group of chemical workers can be reached. The overall mortality of the entire cohort of chemical workers, who were potentially exposed to formaldehyde and other chemicals, was less than (by 26 percent) what one would have expected from a group of U.S. white males similar in age during the same time period. This "healthy worker effect" is quite common in occupational epidemiologic studies and may be the result of several factors: pre-employment selection through physical examinations, self-selection by those who are physically fit to work, and economic stability enjoyed by regularly employed individuals.

To determine whether any of the potential occupational hazards has an adverse health effect, one needs to examine the cause-specific SMRs. For the entire cohort, the observed mortality from cancer of all sites was very similar to the expected. When site-specific cancer mortality was examined, no significant excess was detected. In particular, no nasal cancer death was observed. Mortality analysis by date of hire did not reveal any significant mortality excess either.

Mortality from cancer of the prostate is significantly higher than the expected when a latent period of 20 years is taken into consideration. If employment at the plant was responsible for the excess, one would expect to see an increasing trend with respect to length of employment. However, such a trend was not found. In fact, the excess appears to have come from the group with the shortest employment (<5 years). Furthermore, the ages at death of the 4 cases were 66, 67, 68, and 70. For U.S. white males between 65 and 70, the prostate is one of the most frequent cancer sites. A review of the work histories did not indicate any common exposure, nor any exposure to heavy metal pigments. Since only 4 cases were involved, and there is no indication of a trend by length of employment, this finding must be regarded as preliminary and must be interpreted with caution.

A significant mortality excess in motor vehicle accidents was also detected when mortality experience 20 years after hire was examined. For this particular cause, an occupational etiologic factor is most unlikely. The likelihood of the excess being occupationally related is further diminished by the lack of a trend by length of employment.

Mortality analysis by length of employment reveals an increasing trend between lymphopoietic cancer mortality and length of employment, although none of the individual SMRs in each of the length of employment groups is statistically significant. It should also be emphasized that in four of the five length of employment groups, only one death was involved in each group. A review of the work histories showed that at least three of the six decedents were definitely not exposed to benzene. A thorough analysis by work history or exposure will be performed when the coding of work histories is completed.

In closing, it should be pointed out that there are several limitations in the study, most of which are typical of an historical mortality study on an industrial group. First, although the percent lost to follow-up and the proportion of outstanding death certificates are low, it is possible, but not likely, that deaths from certain rare causes might have been missed. Second, coding of the work histories has not been completed, and analysis by work history or exposure is not feasible at this time. Third, for nasal cancer, the cohort is too small to detect any risk of reasonable magnitude. Fourth, being a mortality study, the investigation not only inherits all the problems associated with death certificates, but also suffers from the lack of in-depth clinical information. Fifth, employment other than that at this particular plant has not been considered. Sixth, it was not possible to include smoking history in the analysis. Seventh, these workers were potentially exposed to a number of chemicals. Analysis by "pure" exposure will not be possible. However, comparisons between high and low exposures to a specific chemical will be made. In spite of these limitations, the study has presented an overall mortality pattern among these chemical workers and has also identified a couple of areas that need further investigation. With regard to formaldehyde, respiratory cancer is the primary concern. The preliminary results presented here demonstrate that the cohort as a whole did not experience any excess in respiratory cancer mortality. A detailed analysis by work history or exposure will be undertaken in the next phase of the study.

REFERENCES

1 Booz, Allen & Hamilton Inc. *Preliminary Study of the Costs of Increased Regulation of Formaldehyde Exposure in the US Workplace.* (February 1979).

2 Chemical Industry Institute of Toxicology. *Progress Report on CIIT Formaldehyde Study.* (January 16, 1980).

3 *International Classification of Diseases*, 1979 Revision. Geneva: World Health Organization (1978).

4 Monson, RR. Analysis of relative survival and proportional mortality. *Comput. Biomed. Res.* 7:325–332 (1974).

5 DeCoufle, P, TL Thomas, and LW Pickle. Comparison of the proportionate mortality ratio and standardized mortality ratio risk measures. *Am. J. Epidemiol.* 111:263–269 (1980).

6 Kilpatrick, SJ. Mortality comparisons in socio-economic groups. *Appl. Statistics* 12:65–86 (1963).

7 Redmond, CK, and PP Breslin. Comparison of methods for assessing occupational hazards. *J. Occup. Med.* 17:313–317 (1975).

APPENDIX: ALGEBRAIC RELATIONSHIP BETWEEN SMR AND PMR

Statistical Notations

	Study group	Comparison group
Number at risk in the jth age group	n_j	N_j
Deaths from the ith cause in the jth age group	d_{ij}	D_{ij}
Total deaths (all causes) in the jth age group	$d_{.j}$	$D_{.j}$

For cause i, PMR is defined as

$$\text{PMR}_i = \frac{\text{Observed deaths due to cause } i \text{ in all age groups}}{\text{Expected deaths based on ratio of specific cause to all causes}}$$

$$= \frac{\Sigma_j d_{ij}}{\Sigma_j [(D_{ij}/D_{.j})d_{.j}]} = \frac{\Sigma_j d_{ij}}{\Sigma_j \{(D_{ij}/N_j)[1/(D_{.j}/N_j)](d_{.j}/n_j)n_j\}}$$

$$= \frac{\Sigma_j d_{ij}}{\Sigma_j \{[(D_{ij}/N_j)n_j][(d_{.j}/n_j)/(D_{.j}/N_j)]\}} = \frac{\Sigma_j d_{ij}}{\Sigma_j [k_j(D_{ij}/N_j)n_j]}$$

where

$$k_j = \frac{d._j/n_j}{D._j/N_j} = \text{age-specific SMR from all causes}$$

If all age-specific SMRs from all causes are equal, i.e., k_j remains constant for all groups ($k_j = k$).

$$\text{PMR}_i = \frac{\Sigma_j\, d_{ij}}{\Sigma_j\, [k(D_{ij}/N_j)n_j]} = \frac{1}{k}\, \frac{\Sigma_j\, d_{ij}}{\Sigma_j\, [(D_{ij}/N_j)n_j]} = \frac{1}{k}\, \text{SMR}_i$$

Furthermore, if one assumes that all age-specific SMRs from all causes equal 1, i.e., $k_j = k = 1$, then $\text{PMR}_i = \text{SMR}_i$.

Part Four

Risk Assessment

Chapter 23

Risk Assessment
with Formaldehyde: Introductory
Remarks

Leon Golberg

The subject of formaldehyde toxicology has by no means been exhausted. Before proceeding to a synthesis of the data for purposes of risk assessment, two topics call for special, albeit brief attention.

What has emerged very clearly from the surveys and epidemiologic studies is the need for some objective, reliable biological indicator of exposure to formaldehyde. As mentioned earlier, formaldehyde is an endogenous intermediary metabolite the rate of formation of which is likely to be increased following demethylation of drugs or food components containing $-NCH_3$, $-OCH_3$, or $-SCH_3$ groups. These facts should not discourage us from searching for one or more indicator compounds present in blood or urine at levels that rise significantly following inhalation or perhaps even percutaneous absorption of formaldehyde. For example, as shown in Fig. 1, endogenously produced formaldehyde forms a number of condensation products with biogenic amines. Such products possess powerful pharmacological properties that qualify them as potential endogenous neuromodulators, perhaps through their actions as false neurotransmitters (1, 2). We are principally concerned with tetrahydroisoquinolines (TIQs) and tryptolines (β-carbolines), the former derived from catecholamines

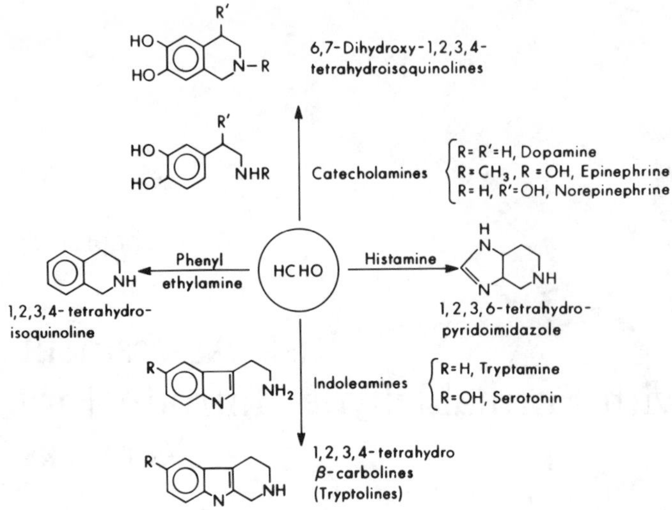

Figure 1 Condensation products of endogenously formed formaldehyde with biogenic amines.

and the latter from indoleamines. In addition to formaldehyde, attention has been focussed on acetaldehyde as a metabolite of ethanol and the role of TIQs and β-carbolines as accounting for the nervous system effects of alcoholism (3, 4).

Ironically enough, 1-methyl-β-carboline (harman) is present naturally in alcoholic drinks such as beer, wine, and saké (5–7), while intraventricular administration of a β-carboline to rats serves to increase the voluntary intake of ethanol (8). Lest we dismiss these observations as curiosities of the scientific literature, let me draw attention to recent reports that ethyl β-carboline-3-carboxylate isolated from human urine and brain has a higher affinity than diazepam for the benzodiazepine receptors in brain (9) and can distinguish between such receptors in the hippocampus and cerebellum (10).

In sum, therefore, we have a series of heterocyclic condensation products produced from formaldehyde. Their presence in mammalian (including human) tissues, blood, and urine has been established. Quantitative methods of measuring these condensation products have been developed in some instances. I would suggest that the work on disposition of radiolabeled formaldehyde within the body be extended in this direction.

The second topic concerns the fact that exposure to formaldehyde so often involves concurrent exposures to a variety of other toxicants with additive, synergistic, or antagonistic actions. Thus in the gas phase of cigarette smoke we are confronted not only by formaldehyde but also by acrolein, acetaldehyde and other aliphatic aldehydes, and oxides of nitrogen (11–13). Ultraviolet irradiation of aqueous formaldehyde gives rise to glyoxal and malonaldehyde (14). What happens to formaldehyde in automobile exhaust emissions, and the nature of the toxicity problems that arise, is another topic to be considered if time and data were available.

Sulfhydryl-containing compounds and ascorbic acid seem to counter the acute effects of aldehydes such as formaldehyde (15, 16), but our principal concern here is with the consequences of long-term, low-level exposures. From the data presented in the previous chapters, the assessment of risk of such exposures, in the tremendous variety of situations of everyday life in which exposures occur, is a substantial undertaking. The chapters that follow are limited to mathematical approaches, to occupationally exposed workers, and to an analysis that seeks to integrate the results of testing and research as a basis for risk assessment. Inevitably, many topics in this area will remain to be considered.

REFERENCES

1   Deitrich, R, and V Erwin. Biogenic amine-aldehyde condensation products: Tetrahydroisoquinolines and tryptolines (β-carbolines). *Ann. Rev. Pharmacol. Toxicol.* 20:55–80 (1980).

2   Buckholtz, NS. Neurobiology of tetrahydro-β-carbolines. *Life Sci.* 27:893–903 (1980).

3   Melchior, CL. Long-lasting effects of tetrahydroisoquinolines. In *Psychopharmacology of Alcohol*, edited by M Sandler, pp. 149–153. New York: Raven Press (1980).

4   Holman, RB, GR Elliott, K Faull, and JD Barchas. Tryptolines: The role of indoleamine-aldehyde condensation products in the effects of alcohol. In *Psychopharmacology of Alcohol*, edited by M Sandler, pp. 155–169, New York: Raven Press (1980).

5   Lashkhi, AD, and LA Mudzhiri. Alkaloid harman in wine. *Vinodel. Vinograd. SSSR.* 1:53 (1975).

6   Takase, S, and H Murakami. Fluorescence of sake. I. Fluorescence spectrum of sake and identification of harman. *Agr. Biol. Chem. (Tokyo)* 30:869 (1966).

7   Beck, O, and B Holmstedt. Analysis of 1-methyl-1,2,3,4-tetrahydro-β-carboline in alcoholic beverages. *Fd. Cosmetic Toxicol.* 19:173–177 (1981).

8   Melchior, CL, and RD Myers. Alcohol drinking induced in the rat after chronic injections of tetrahydropapaveroline (THP), salsolinol or noreleagnine in the brain. In *Alcohol and Aldehyde Metabolizing Systems*, edited by RG Thurman, JR Williamson, HR Drott, and B Chance, vol. III, p. 545. New York: Academic Press (1977).

9   Braestrup, C, M Nielsen, and CE Olsen. Urinary and brain β-carboline-3-carboxylates as potent inhibitors of brain benzodiazepine receptors. *Proc. Natl. Acad. Sci. USA* 77:2288–2292 (1980).

10 Nielsen, M, and C Braestrup. Ethyl β-carboline-3-carboxylate shows differential benzodiazepine receptor interaction. *Nature* 286:606–607 (1980).

11 *The Health Consequences of Smoking, a Report to the Surgeon General.* U.S. Dept. Health, Education, and Welfare Publication No. (HSM) 72-7516 (1972).

12 Watanabe, T, and DM Aviado. Functional and biochemical effects on the lung following inhalation of cigarette smoke and constituents. II. Skatole, acrolein, and acetaldehyde. *Toxicol. Appl. Pharmacol.* 30:201–209 (1974).

13 Egle, JL, Jr., and PM Hudgins. Dose-dependent sympathomimetic and cardioinhibitory effects of acrolein and formaldehyde in the anesthetized rat. *Toxicol. Appl. Pharmacol.* 28:358–366 (1974).

14 Halmann, M, and S Bloch. Glyoxal and malonaldehyde formation by ultraviolet irradiation of aqueous formaldehyde. *BioSystems* 11:227–232 (1979).

15 Sprince, H, CM Parker, and GG Smith. Comparison of protection by L-ascorbic acid, L-cysteine, and adrenergic-blocking agents against acetaldehyde, acrolein, and formaldehyde toxicity: Implications in smoking. *Agents and Actions* 9:407–414 (1979).

16 Guerri, C, W Godfrey, and S Grisolia. Protection against toxic effects of formaldehyde *in vitro*, and of methanol or formaldehyde *in vivo* by subsequent administration of SH reagents. *Physiol. Chem. Physics* 6:543–550 (1976).

Chapter 24

Mathematical Approaches to Risk Assessment: Squamous Cell Nasal Carcinoma in Rats Exposed to Formaldehyde Vapor

David W. Gaylor

Nasal tumors were reported by Swenberg et al. (1) in Fischer 344 rats exposed to formaldehyde vapor by inhalation. Animals were exposed to average dosages of formaldehyde vapor of 0, 2.1, 5.6, and 14.1 ppm, 6 h/day, 5 days/week, for 24 months. Results were reported only for the first 18 months. A total of 120 animals per sex were exposed at each dosage level. Within each group, 10 animals were sacrificed at 6 months, 10 animals at 12 months, and 20 animals at 18 months. Since time to tumor data and cause of death were not presented, risk analyses can be performed only on the unadjusted tumor rates of animals dying with tumors through 18 months. Hence, competing risk analyses utilizing the sacrificed animals cannot be performed. Thus, there are 100 animals per sex per dosage that were not scheduled for an interim sacrifice prior to 18 months. Since no significant differences were noted in the frequency of nasal tumors between sexes during the first 18 months of exposure, the results for the sexes were combined by Swenberg et al. (1). The proportions of animals dying *with* squamous cell carcinomas before 18 months were 0/200, 0/200, 0/200, and 28/200 for 0, 2.1, 5.6, and 14.1 ppm, respectively. Note that these animals are dying *with* squamous cell carcinomas and may or may not be dying *due* to the carcinomas.

A significant proportion of animals exhibited squamous cell carcinomas at 14.1 ppm, and none was observed at 5.6 ppm or below during the first 18 months. The problem addressed here is to predict the level of cancer risk that might occur in rats at formaldehyde vapor levels that occur in home and occupational environments.

SAFETY FACTORS

The largest "no observable effect level" (NOEL) for squamous cell nasal carcinoma is 5.6 ppm. Since formaldehyde produced carcinomas at a higher dosage, a relatively large safety factor is required. If a safety factor of 1000 is applied to the largest NOEL, an allowable dosage of 5.6 ppm ÷ 1000 = 0.0056 ppm or 5.6 ppb is obtained. The problem is to determine how much protection this safety factor provides.

Since the dose response is apparently at least curving upward, cancer risk decreases faster than dosage as dosage is lowered. Hence, decreasing dosage by a safety factor of F will apparently decrease cancer risk by more than a factor of F. With no animals in 200 exhibiting cancers at 5.6 ppm, the upper 95 percent confidence limit on the proportion of rats with cancer is 0.015: that is, if the true cancer rate at 5.6 ppm were 1.5 percent, there is still a 5 percent chance that in 200 animals there would be no tumors observed. If the risk (proportion of rats with squamous cell nasal carcinomas) at 5.6 ppm may be as high as 0.015, then the risk at one-thousandth that dosage is predicted to be less than one-thousandth the risk at 5.6 ppm. Thus, the risk at 5.6 ppb of formaldehyde is predicted to be between 0 and $0.015 \div 1000 = 15 \times 10^{-6}$ or 15 per million animals.

If a lower upper limit of risk is desired, then a larger safety factor must be used. If it is desired to limit the risk to 10^{-6}, then a safety factor of 15,000 is required. Then, the allowable dosage becomes 5.6 ppm ÷ 15,000 = 0.00037 ppm or 0.37 ppb of formaldehyde in the air.

In general, the excess risk (R) above background at the NOEL $\div F$ dosage is less than the upper confidence limit (L) on excess risk at the NOEL divided by the safety factor (F). That is, the excess risk at the NOEL $\div F$ dosage is less than $R = L/F$. Conversely, the size of the safety factor required for a given level of excess risk is obtained by $F = L/R$. Hence, the size of the safety factor can be reduced by reducing the upper confidence limit (L), which can be accomplished by increasing the number of animals on test.

MATHEMATICAL MODELS FOR LOW DOSE EXTRAPOLATION

Because of limited resources, it is necessary to conduct animal bioassays at dosage levels above human exposure levels in order to detect potential toxic effects with relatively small numbers of animals. Hence, the problem arises of extrapolating results at experimental dosage levels to lower levels experienced by humans. No attempt will be made here to consider the problem of extrapolation from animals to humans. This discussion is limited to predicting risk within the animal population.

Various mathematical models have been proposed for low dosage extrapolation. Among these are the probit, logistic, Weibull, extreme value, one-hit, multihit, gamma multihit, and multistage. Generally, these models fit experimental data about equally well and are indistinguishable from each other in the experimental dose range. However, point (best) estimates of risk at low doses often differ by orders of magnitude for the various models (2). Since, there is no mathematical or biological basis for selecting a model, point (best) estimates of risk at low doses are meaningless. However, if the true dose response is curving upward at low doses, it is possible to obtain an upper limit on the risk at low doses. In the low dose portion, wherein the response is curving upward, a line from a point in this portion of the curve connected to the spontaneous background response at zero dose will always lie above the true dose-response curve. Hence, points along this line provide an upper limit on the true risk.

The linear algorithm proposed by Gaylor and Kodell (3) is used here to estimate risk limits at low doses below the experimental dose range. The first step is to select any mathematical model that adequately fits the data in the experimental dose range. The purpose of fitting a model is to use all of the data in order to obtain an upper confidence limit on the risk at the lowest experimental dose that reflects the shape of the dose-response curve. Then a straight line from the origin to the upper confidence limit at the lowest experimental dosage provides an upper limit of risk on dosages below the experimental dosage range. If L is the upper confidence limit on the excess risk above background at the lowest experimental dosage (D_E), then the upper limit on the excess risk (R) at a low dosage (D) is $R = D \times L/D_E$.

A multistage model (4) was fit to the proportion of animals dying with squamous cell nasal carcinomas before 18 months, as reported by Swenberg et al. (1). The upper 95 percent confidence limit on the excess risk is 0.0030 at 2.1 ppm. Hence, the upper limit on the excess risk above background of squamous cell nasal carcinomas in Fischer 344 rats exposed to a dosage of D ppm of formaldehyde for 6 h/day, 5 days/week for 18 months is

$$R = D \times 0.0030/2.1 = 0.0014\,D$$

Thus, the excess carcinoma rate at 1 ppm is predicted to be between 0 and 0.0014, i.e. less than 1.4 per 1000 animals. The linear interpolation algorithm is shown graphically in Fig. 1. The dosage corresponding to a maximum risk of 10^{-6} is $D = 10^{-6}/0.0014 = 0.0007$ ppm or 0.7 ppb. Recall, that the use of a safety factor on the largest NOEL required a lower dosage, 0.37 ppb, to provide

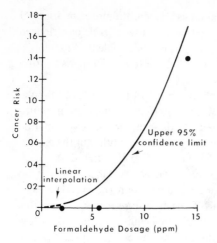

Figure 1 Potential proportion of Fischer 344 rats dying with squamous cell nasal carcinomas before 18 months.

the same risk limit. The linear algorithm technique allows nearly double this dosage because the linear algorithm technique makes more effective use of the experimental data.

If the upper confidence limit on the risk at 2.1 ppm were only based on the information that no animals in 200 developed carcinomas at that dosage, the upper 95 percent confidence limit would have been 0.015. Thus, the upper confidence limit at 2.1 ppm was reduced to 0.003 by fitting a curve. Hence, the upper limit on risk at lower dosages is reduced by a factor of 0.015/0.003 = 5 due to utilizing a curve in the data range. In general, the use of linear interpolation is less conservative than the use of safety factors.

DISCUSSION

Since the linear interpolation algorithm does not extrapolate to doses below the experimental range by using a mathematical model, the unresolvable issue of selecting a mathematical model is circumvented. Hence, selection of the mathematical model used in the experimental range is not a critical issue, provided a valid estimate of the upper confidence limit is obtained at the lowest experimental dosage from which interpolation proceeds.

The linear interpolation algorithm makes use of all the experimental data and the curvature of the dose response down to the lowest experimental dosage. Therefore, it is generally less conservative, i.e., predicts lower risks resulting in higher allowable dosages, than the application of a safety factor to a "no observable effect level" (NOEL). The NOEL also suffers because it has no precise meaning since it is a statistical quantity that depends on the number of animals tested.

It is often noted that risk analyses predict allowable dosages so low that they are not attainable. This is not a fault of the mathematical techniques used but simply is a limitation of the animal bioassay technique. It is not possible to measure precisely biological effects that occur with a low incidence rate. Thus, it is not possible to insure a high degree of safety from a test on a small number of animals, particularly if tumors are obtained in animals at dose levels not far above human exposure levels, as is the case with formaldehyde.

REFERENCES

1 Swenberg, JA, WD Kerns, RI Mitchell, EJ Gralla, and KL Pavkov. Induction of squamous cell carcinomas of the rat nasal cavity by inhalation exposure to formaldehyde vapor. *Cancer Res.* 40:3398–3402 (1980).

2 FDA Advisory Committee on Protocols for Safety Evaluation. Panel on carcinogenesis report on cancer testing in the safety evaluation of food additives and pesticides. *Toxicol. Appl. Pharmacol.* 20:419–438 (1971).

3 Gaylor, DW, and RL Kodell. Linear interpolation algorithm for low dose risk assessment of toxic substances. *J. Environ. Pathol. Toxicol.* 4:305–312 (1980).

4 Crump, KS, HA Guess, and KL Deal. Confidence intervals and tests of hypotheses concerning dose response relations inferred from animal carcinogenicity data. *Biometrics* 33:437–451 (1977).

Chapter 25

Risk Assessment for Exposure to Formaldehyde

John J. Clary

This approach to risk assessment as it relates to formaldehyde exposure will attempt to define the various risks that are known or predictable from either human or animal data and will be non-mathematical. Also briefly discussed are the dose-response relationships, species differences, and the difficult task of extrapolating from animals to humans.

TYPES OF RISK

The potential risks from formaldehyde exposure that are derived from human experience are related primarily to dermal and respiratory effects—irritation and sensitization. The human data can be used to establish dose-response relationship in this area.

Animal data are suggestive of a risk in other areas. One area is a mutagenic response that involves an in vivo or an in vitro reaction of formaldehyde with a genetic material. Any potential problem in this area is likely to appear in future generations. Another concern is the potential risk from cancer. Both of these areas, of course, are of great concern and the extrapolation from animals to humans is extremely important and difficult.

Lastly there are some suggestions in the literature of the potential for other problems to be associated with formaldehyde. None of these appears to be well enough documented or studied to be a defined risk.

POPULATIONS AT RISK

The populations that are potentially exposed to formaldehyde include the general population, the occupational group, and a special population. The general public is exposed to formaldehyde basically in two ways: a) exposure to natural products of formaldehyde, e.g., from generation of formaldehyde as a result of such natural reactions as combustion; b) exposure to consumer products that contain formaldehyde—cosmetics, permanent press fabric, toothpaste, urea-formaldehyde foam, and particleboard. This area was covered in detail in Chap. 1.

The occupational group is comprised of both the producer and the user of formaldehyde. Because of the many uses of formaldehyde, the user population and its exposure is diverse.

The special population is difficult to define, but it contains all people who do not respond to formaldehyde in a normal fashion. This special population can also be subdivided into two groups at risk: hypersensitive and sensitized. The hypersensitive individual reacts at lower levels of exposure than the normal population. This special population could conceivably also include people with respiratory problems and perhaps the very young. Hypersensitization is characterized as follows:

1 Response is a normal expected effect of a chemical.
2 Response does not require preconditioning of individual.
3 Reaction occurs following the dose of a compound that is smaller than the usual dose.

The sensitized individual responds atypically to formaldehyde in that the response is mediated by an immune mechanism. Sensitization usually is not dose related and frequently involves only a small fraction of the total exposed population. Pre-exposure to the chemical is required to produce a toxic effect via an antibody.

KNOWN HUMAN RISKS

When we look at the risks themselves we are talking about respiratory irritation, which is sensory irritation—the burning, tearing, and stinging of the eye, nose, or throat. Many studies have correlated airborne levels with subjective symptoms of sensory irritation. Essentially, the NIOSH criteria document (1) and, of course, the later National Academy of Science document (2) show a wide range of airborne levels that reportedly cause the symptoms noted in Table 1.

Table 1   Airborne Formaldehyde and Sensory Irritation in Humans

Concentration (ppm HCHO)	Duration of exposure	Responses
1–11	8 h/day	Eye, nose, throat irritation
13.8	30 min	Nose and eye irritation subsiding after 10 min in chamber
0.13–0.45	?	Complaints of temporary eye and upper respiratory tract irritation
16–30	8 h/day	Eye and throat irritation, skin reaction
0.9–1.6	8 h/day	Itching eyes, dry and sore throats, disturbed sleep, unusual thirst on awakening in the morning
0.3–2.7	8 h/day	Annoying odor, constant prickling irritation of the mucous membranes, disturbed sleep, thirst, heavy tearing
0.09–5.26 (with para-formaldehyde)	1 h	Eye and upper respiratory irritation, lessened during the day
0.9–3.3	1 h	Mild eye irritation, objectionable odor
0.9–2.7	1 h	Tearing of eyes, irritation of nasal passages and throat (irritant effects greatest at very beginning of workday and after lunch)
2.1–8.9 0.5–3.3	Daily	Increased occurrence of upper respiratory irritation
3	?	Irritation of the conjunctivas, nasopharynx, skin

Relatively low and high airborne levels of formaldehyde are reported in these studies. Basically, however, the data suggest that the sensory irritation in a significant number of the occupational population is first seen at levels around 1 ppm. Because of analytical reasons, some of these airborne levels are subject to question.

If the normal population is similar to the occupational group then we can define the threshold effect level for sensory irritation to formaldehyde also at levels around 1 ppm. This sensory response could be considered a warning of the presence of formaldehyde (with tearing and nasal discharge associated with this response as a mechanism of removing formaldehyde from sensitive body surfaces, i.e., a defense mechanism).

Little information on respiratory sensitization is available. This sensitization is a true immunological response, not a hypersensitive response. In this type of response it is necessary to have pre-exposure to the chemical "formaldehyde" before this antibody-mediated effect is produced. The only paper of any merit was published by Hendricks and Lane (3). It reports case studies of hospital workers who used formaldehyde to clean dialysis machines. When these workers were challenged by formaldehyde vapor, some responded with asthmatic-type symptoms. There is no good information on the rate of true respiratory sensitization in an occupational or nonoccupational group. The published evidence suggests, however, that it is very small, much less than 0.1 percent.

Another response is irritation, or redness, erythema, and inflammations on the skin resulting from exposure to solutions of formaldehyde. Most data on dermal response are derived from the occupational group. We know that formaldehyde is a skin irritant and an aqueous solution of formaldehyde above 2 percent will elicit an irritant response in the majority of the population. There is also a small potential for dermal irritation from nonoccupational exposures, especially in the hypersensitive individual. However, technology has greatly reduced the availability of free formaldehyde in consumer products, and thus a dermal response is unlikely in the normal population. Information on dermal sensitization is better than that on respiratory sensitization. Up to 1-4 percent of the population can become dermally sensitized to formaldehyde—true sensitization according to Maibach (see Chap. 15, this volume). In a nonoccupational setting the information is less clear, but sensitization is possible if there is significant dermal contact.

Several important points should be made about human effects. First, the risks are, in most cases, transitory in nature—the irritation effect being reversible. Second, formaldehyde has a good warning property: it has its own built-in alarm system that warns individuals when they are being exposed to an excess amount. The human data suggest that a threshold concentration is important in defining the risk of these types of reversible responses in the normal population.

TESTS OF BIOLOGICAL ACTIVITY

Mutagenicity (Table 2)

The mutagenicity test results are divided into two categories: in vitro and in vivo. Formaldehyde shows activity in many in vitro tests. The Ames test, which is reported to have the best correlation of any of these tests between the carcinogenicity and mutagenicity, is negative for formaldehyde. One possible explanation might be that formaldehyde is lethal to the bacteria and at a nontoxic dose the mutagen rate may be much too low to be picked up by this type of test. Again, sometimes the lethality of a material becomes more important from a risk assessment point of view than mutagenicity, which might not be expressed except at levels above the lethal level for animals or humans. There are only two in vivo studies reported in the literature (1) in this area. One is the *Drosophila*, which is positive, and the other is a mammalian in vivo test (the dominant lethal), which is negative. Brusick reported a positive finding for

**Table 2   Short-Term Tests for Mutagenicity
of Formaldehyde**

Genetic activity detected	No genetic activity detected
In vitro	
E. coli	Salmonella
Saccharomyces	Chromosomal aberration
Mouse lymphoma	
Chinese hamster ovary	
Sister chromatid exchange	
Unscheduled DNA repair	
Neurospora	
Aspergillus	
In vivo	
Drosophila	Dominant lethal
Sister chromatid exchange	

the sister chromatid exchange following in vivo exposure (see Chap. 8, this volume). The Formaldehyde Institute studies in this area should help give us an additional understanding of the importance of the in vivo response in mammalian systems. I think this study could be very important in defining the mutagenic risk to humans.

Carcinogenicity

There are many things to consider about the carcinogenic response—questions of irritation, species difference, thresholds. Key issues relating to formaldehyde risk assessment in this area are as follows:

> Epigenetic versus genetic
> Metaplasia
> Species difference
> Concentration versus total dose
> Respiratory pattern
> Metabolism
> Extrapolation to humans

The species response provides some very clear-cut information (Table 3). The CIIT study report (4) shows a carcinogenic response as well as hyperplasia and metaplasia in rats exposed to 15 ppm and two to three compound-related cancers at 6 ppm. At lower study levels (2 ppm) only metaplasia was observed in the rats in the CIIT study. Two other studies use rats. One was a poorly done study in Japan in which 4 of 10 rats developed sarcomas (5) following subcutaneous injections of formaldehyde. There were no controls in this study and the route of exposure is unrealistic. In the second study (6), rats were exposed to a mixture of formaldehyde and hydrogen chloride, which theoretically could form bis(chloromethyl)ether in the atmosphere. The response seen in this study shows several tumor types, including the squamous cell carcinoma that was seen in the CIIT study at 15 ppm. Because it was a mixed exposure, it is not possible to attribute the response to only formaldehyde.

The mouse in the CIIT study showed a very weak carcinogenic response to formaldehyde at 15 ppm. The only other study using mice was done by Horton et al. (7) at approximately 40, 80, and 120 ppm formaldehyde exposure for 35–64 weeks. The mice were exposed only three times a week, 1 h per exposure. This study showed no excessive tumor formation. However, exposure times and numbers of exposures per week were insufficient.

As far as hamsters are concerned, the only information on response in this third species is work at the Oak Ridge National Laboratory (Nettesheim, D, unpublished data, 1975–1977). This study looked at formaldehyde and other materials as potential carcinogens. One group of animals in the Oak Ridge studies, however, was exposed to 10 ppm formaldehyde for its lifetime. These data suggest that formaldehyde itself was not carcinogenic in the hamster.

Table 3 Carcinogenicity Studies of Formaldehyde

Species	Route	Response	Comment
1. Rat	Subcutaneous	4 of 10 with sarcomas	No control
2. Rat	Inhalation 6 h/day, 5 days/wk for 1 yr at 0, 2, 6, and 15 ppm	1. Squamous cell carcinoma and hyperplasia in nasal cavity at 15 ppm and 6 ppm 2. Squamous metaplasia at 2 ppm and 6 ppm	
3. Mouse	Inhalation 1 h/day, 3 days/wk for 35–64 wk at 0, 40, 80, and 120 ppm	1. No increase in tumors 2. Tracheal hyperplasia and metaplasia observed at level above 80 ppm	1. Exposure level very high 2. Duration of exposure short
4. Mouse	Inhalation 6 h/day, 5 days/wk for 2 yr at 0, 2, 6, and 15 ppm	1. Very slight increase in tumors at 15 ppm 2. Squamous cell metaplasia at 15 ppm	
5. Hamster	Inhalation 5 times a week for lifetime at 10 ppm	1. No increase in tumors 2. Not a cocarcinogen	Study has not been published; details lacking; 88 animals only

A fourth species should be added to this list. In the three epidemiologic studies to date there is no indication of any increase in cancer in any area of the respiratory tract in humans. The study of morticians presented by Walrath and Fraumeni (see Chap. 20, this volume) suggested some increase in skin cancer, but the evidence to link this find to formaldehyde is very weak. The preliminary results of the CIIT study raise the question of mechanism. This question is important in risk assessment. Is the mechanism of action epigenetic or genetic? In the genetic mechanism the potential carcinogens react with the sites on the bases of DNA to disturb normal base pairing. If these altered bases are not removed by normal DNA repair, they may become permanent genetic changes. These types of changes are manifested after a long latent period in tumor formation. For an epigenetic or cytotoxic agent, tissue damage occurs to such an extent as to cause cell death followed by subsequent cell regeneration. Swenberg et al. (see Chap. 12, this volume) reported a 70-fold increase in cell proliferation in the nasal cavity of rats exposed to 15 ppm formaldehyde. Prolonged chronic administration of an epigenetic or cytotoxic agent results in stimulation of DNA replication in the affected tissue. Since there is always a chance for a mistake in the replication cycle, an epigenetic or cytotoxic agent can cause a spontaneous increased rate of mutations. When we consider DNA repair and replication in this mechanism, the DNA repair system may become less effective as cell replacement is stimulated. This causes many more DNA changes, i.e., mutagenic response and this can actually be expressed later on as tumors. The epigenetic reaction is due not to the reaction of the chemical with the DNA but to the overload of the normal DNA repair mechanism.

The important difference in epigenetic versus genetic mechanisms is that in a genetic mechanism there is a potential, no matter how small, for response at any exposure level. An epigenetic mechanism, however, implies a threshold level below which there is no response, i.e., tissue damage or cell death resulting in rapid cell proliferation and overloading of the normal DNA repair mechanism. The CIIT data suggest a threshold for tumor formation. Threshold levels for epigenetic agents that would not produce tissue damage in either animals or humans can be suggested and supported by good industrial hygiene practices and good toxicological data. These levels, therefore, could be used to establish an acceptable workplace level. The CIIT results point out that formaldehyde induces squamous cell metaplasia. The production of squamous cell metaplasia varies from species to species, depending on exposure. It should be considered a more severe compound-related effect than just sensory irritation, but it can still be considered a protective mechanism. Certainly if formaldehyde reacts according to an epigenetic mechanism, production of the metaplasia is extremely important for controlling cell death and DNA repair. Table 4 shows additional data on metaplasia on different species. In the rats, CIIT reported metaplasia at levels as low as 2 ppm (4). CIIT has also reported separately that metaplasia can be produced in 1 week at 15 ppm. This suggests that 15 ppm

Table 4 Effect of Airborne Concentration of Formaldehyde on the Formation of Respiratory Tract Metaplasia

Species	Concentration (ppm)	Time	Response
Rat	1	7 days/wk, 22 h/day for 6 mon	No metaplasia
	2	5 days/wk, 6 h/day for 18 mon	Metaplasia—nasal cavity
	3	7 days/wk, 22 h/day for 6 mon	Metaplasia—nasal cavity
	6	5 days/wk, 6 h/day for 18 mon	Metaplasia—nasal cavity
	15	5 days/wk, 6 h/day for 6 mon	Metaplasia—nasal cavity
Mouse	2	5 days/wk, 6 h/day for 18 mon	No metaplasia
	6	5 days/wk, 6 h/day for 18 mon	Metaplasia—nasal cavity
	15	5 days/wk, 6 h/day for 18 mon	Metaplasia—nasal cavity
	40	3 days/wk, 1 h/day for 35 wk	No metaplasia
	80	3 days/wk, 1 h/day for 35 wk	Metaplasia—tracheobronchial
	120	3 days/wk, 1 h/day for 35 wk	Metaplasia—tracheobronchial
Hamster	1	7 days/wk, 22 h/day for 6 mon	No metaplasia
	3	7 days/wk, 22 h/day for 6 mon	No metaplasia
	10	5 days/wk for lifetime	No information
	250	1 h/day up to 15 days	Metaplasia—tracheobronchial
Monkey	1	7 days/wk, 22 h/day for 6 mon	No metaplasia
	3	7 days/wk, 22 h/day for 6 mon	Metaplasia—nasal cavity

may be too high a level for long-term exposure in the rat. In the Formaldehyde Institute 6-month inhalation study, the rats exposed at 3 ppm developed squamous cell metaplasia (Formaldehyde Institute, unpublished data, 1980). Over the 6-month period, however, the animals were exposed for 7 days/week, 22 h/day. This was about five times as long per week as exposures in the CIIT study.

A total dose can be considered by multiplying concentration by time. At 3 ppm in the Formaldehyde Institute study, this would be equivalent on a total dose basis to 15 ppm at 6 months in the CIIT study. In the Formaldehyde Institute study, however, metaplasia was not observed in rats exposed at 1 ppm. Again, using the same logic, 1 ppm for 22 h/day, 7 days/week for 6 months would give an equivalent total dose equal to 5 ppm if exposure was only 6 h/day, 5 days/week for 6 months. Metaplasia was seen at 2 ppm in the CIIT study. This suggests that what we are looking at is not a total dose (concentration × time) but is strictly one of concentration or threshold. Based on the CIIT data and the Formaldehyde Institute data, 1 ppm does not appear to cause metaplasia, but 2 ppm and above does. This response, of course, is seen in a very sensitive species, the rat.

CIIT data for a second species, the mouse, shows metaplasia only at 18 months at 15 ppm—quite different from the rat, which shows metaplasia in one week following exposure to formaldehyde at 15 ppm. In an inhalation study in mice, tracheal and bronchial metaplasia were reported by Horton et al. (7).

No indication of any nasal metaplasia was reported, but the paper does not indicate that the nasal cavity was examined.

The hamster, a third species in the Formaldehyde Institute study, had no metaplasia at 3 ppm, suggesting a species difference between rat and hamster.

In the Oak Ridge study, no metaplasia was reported in the nasal cavity of hamsters exposed at 10 ppm.

An interesting paper by Schreiber et al. (8), which compared the tracheal metaplasia induced by benzopyrene or formaldehyde following high level, short exposures, shows some interesting observations. The animals exposed to formaldehyde showed a quick reversal of the metaplasia within a couple of weeks after being removed from exposure. This suggests that this type of metaplasia, although very similar morphologically to the type seen with benzopyrene, was readily reversible. In contrast, however, the benzopyrene-induced metaplasia progressed even after the exposure was terminated, suggesting a difference in response between these two materials. The authors concluded that the formaldehyde-induced metaplasia is just an irritant response.

Responses in two other species—monkeys and humans—are also important. The data on the monkey in the Formaldehyde Institute study exposure at 3 ppm showed squamous metaplasia in the nasal cavity. There were also indications of rhinitis. The monkey may be a good model for humans. First, rodents are obligatory nose breathers, whereas monkeys and humans are not, i.e., they will breathe through the mouth if they are severely irritated by material. This might even suggest some area other than the nose that would be of concern in monkeys and humans. In the Formaldehyde Institute study, no compound-related response was seen except in the nasal cavity following exposure at 3 ppm for 6 months. This study is the only data we have on formaldehyde's effect in the nonobligatory-nose-breathing animal.

Another point to consider in comparing humans and monkeys is the species response to methanol. Methanol is reported to be metabolized to formaldehyde and then to formic acid (9). Among the animal models that most closely mimic humans in their response to methanol poisoning, only the monkey became blind following methanol exposure. This blindness in monkeys and humans appears to be correlated to the formation of formic acid and the resulting acidosis. Thus the rat and monkey metabolize methanol via the same pathways and produce similar metabolic end-products, but the rate of metabolism may differ significantly between these two species.

Reviewing the information on metaplasia shows that the rat is probably the most sensitive species. If we consider an epigenetic event as wholly or partly responsible for the effect seen in the CIIT study, one has to wonder about the sensitivity of the rat and the role of irritation, cell death, and DNA repair in risk assessment. The CIIT data should also be examined from the other point of view, i.e., is the formaldehyde response a genetic mechanism? The mutagen studies that were previously discussed to indicate that formaldehyde is capable of causing mutations in test systems. This suggests a genetic mechanism. The CIIT cell transformation data in the 10 $T\frac{1}{2}$ cell line suggest formaldehyde is an initiator in the system. Considering all the factors in the CIIT study, such as species difference and sensitivity, raises the question of the mechanism for the response at 15 ppm in the rat—whether this is primarily due to severe irritation, i.e., whether formaldehyde's primary action is an epigenetic agent or a promoter, as suggested by Swenberg et al. (Chap. 12, this volume). The CIIT response in rats can be explained by two different mechanisms. The promoter or epigenetic effect seems to be the predominant of the two mechanisms.

CONCLUSION

Table 5 shows that assessment of the risk associated with formaldehyde involves considering the different populations potentially exposed and the potential health effects associated with each risk. Both animal and human data are being used to evaluate the risk to humans of formaldehyde exposure. This risk assessment covers all phenomena from sensory irritation to the potential for cancer. There appears to be some correlation between the responses in different species and concentrations. A majority of humans respond to the sensory irritation of formaldehyde at approximately 1 ppm. The next level of response is tissue damage as a result of exposure beyond the level that produces sensory irritation. This response is seen in several animal species, the rat being the most sensitive. The final event following high level exposure is cancer in one animal species.

Risk assessment of formaldehyde should also consider its built-in warning properties, species difference, and the problem of extrapolating from animals to humans. Obviously more information is needed. The CIIT study and other studies, especially epidemiology studies, such as the NCI-FI study are not finished. More information on the mechanism of action, species difference, and concentration at target sites, particularly as they relate to carcinogenicity, is needed before the effect of formaldehyde on human health can be accurately predicted.

Table 5 Formaldehyde Thresholds

Response	Level	Species
Odor	below 1 ppm	Humans
Sensory irritation	around 1 ppm	Humans
Metaplasia	2 ppm	Rats
Hyperplasia and squamous cell carcinoma	6–15 ppm	Rats

REFERENCES

1 *Criteria for a Recommended Standard Occupational Exposure to Formaldehyde.* Cincinnatti: National Institute for Occupational Safety and Health, USPHS (1976).

2 Board on Toxicology and Environmental Health Hazard. *Formaldehyde—An Assessment of its Health Effects.* Washington, D.C.: National Academy of Sciences (1980).

3 Hendrick, SD, and PJ Lane. Occupational formalin asthma. *Br. J. Ind. Med.* 24:11–18 (1977).

4 Swenberg, JA, WD Kern, RI Mitchell, EJ Gralla, and KL Pavkov. Induction of squamous cell carcinomas of rat nasal cavity by inhalation exposure to formaldehyde vapor. *Cancer Res.* 40:3398–3402 (1980).

5 Watanabe, F, T Matsanage, T Soejima, and Y Iwata. Study on the carcinogenicity of aldehyde. *Gann* 45:451–452 (1954).

6 Rusch, GM, AR Sellekumar, SL LaMendola, GV Katz, S Laskin, and RE Albert. Inhalation studies with combined formaldehyde and hydrogen chloride vapors. (submitted for publication).

7 Horton, AW, T Russell, and KL Stemmer. Experimental carcinogenesis of the lung and inhalation of gaseous formaldehyde or aerosol of coal tar by C3H mice. *J. Natl. Cancer Inst.* 30:31–43 (1963).

8 Schreiber, H, M Bibbo, GL Wied, G Saccomanno, and P Nettesheim. Bronchial metaplasia as a benign or premalignant lesion. *Acta Cytol.* 23:496–503 (1979).

9 McMartin, KE, G Martin-Amat, PE Noker, and TR Tephly. Lack of a role for formaldehyde in methanol poisoning in the monkey. *Biochem. Pharmacol.* 28:643–649 (1979).

Chapter 26

Risk Assessment Using a Combination of Testing and Research Results

James E. Gibson

Formaldehyde is an important endogenous biochemical occurring in most life forms. As a commodity chemical, formaldehyde is in commonplace use for the production of materials important in modern society. Until recently, very little was known about the potential chronic toxicity and carcinogenicity of formaldehyde.

Control of human exposure to formaldehyde is desirable to avoid well-known irritation effects of the skin, eye, nose, and throat. However, because of physiological defense mechanisms, humans do not normally subject themselves to irritating concentrations of this chemical. Additionally, since formaldehyde occurs normally within cells, human exposure to slight amounts of formaldehyde may be assimilated without toxic effects. Carcinogenicity in rats due to inhalation of high concentrations of formaldehyde for long periods of time has, however, raised additional concerns for human exposure.

The author is grateful to Drs. J. Kovar and D. Krewski, Health Protection Branch, Health and Welfare Canada, and to Dr. T. Starr and Mr. R. Buck, Chemical Industry Institute of Toxicology, for their assistance with the risk assessment analyses.

The Chemical Industry Institute of Toxicology (CIIT) sponsored a long-term inhalation toxicity and carcinogenicity study of formaldehyde vapor in equal numbers of male and female Fischer-344 rats and B6C3F1 mice with exposure at 6 h/day, 5 days/week. During 24 months of inhalation exposure, approximately one half of the rats exposed to the highest exposure concentration, 15 ppm, developed squamous cell carcinomas of the nasal cavity (Tables 1 and 2) or, to a much lesser extent, other tumors of the nasal cavity (see Kerns et al., Chap. 11, this volume). In rats exposed to 15 ppm formaldehyde, the earliest tumors were discovered between the eleventh and twelfth month of exposure. The incidence had sharply increased by the 24th month. In contrast, mice were substantially resistant to these effects (Tables 1 and 2). During examination of tissues taken at the planned necropsy, two squamous cell carcinomas were discovered in mice exposed to 15 ppm only after the 24th month of exposure was completed. Similarly two squamous cell carcinomas were found in the nasal cavities of rats exposed to 6 ppm formaldehyde for 24 months. Thus the latency period for tumor formation as well as the sensitivity to tumor formation was species and dose dependent. The existence of squamous cell carcinoma in the nasal cavities of rats and mice is exceedingly rare. None has been found in approximately 2000 control animals studied by CIIT.

From the above it is clear that the assessment of risk for formaldehyde exposure requires a detailed understanding of the mechanisms involved in the observed neoplastic response. Prediction of risk due to formaldehyde exposure of mice using only the rat data, or vice versa, without such mechanistic information is difficult enough. Extrapolations to humans, however, would require even more assumptions.

Research undertaken by CIIT has provided some of the necessary knowledge for interpreting the results of the long-term studies. The carcinogenic response of rats to high concentration, long-term exposure to formaldehyde is mechanistically complex and multistage (Fig. 1).

Table 1   Formaldehyde Inhalation for 24 Months[*]: Incidence of Squamous Cell Carcinoma of Nasal Cavity

Formaldehyde concentration (ppm)[†]	Number of tumors/nasal cavities examined	
	Rat	Mouse
0	0/232 (0%)	0/223 (0%)
2	0/236 (0%)	0/214 (0%)
6	2/235 (0.9%)	0/218 (0%)
15	103/232 (44.4%)	2/225 (0.9%)

[*]6 h/day, 5 days/wk. The study was initiated with 240 Fischer-344 rats and 240 B6C3F1 mice, evenly divided by sex.

[†]Target concentrations. Actual average concentrations were 0, 2.0, 5.6, and 14.3 ppm.

Table 2 Formaldehyde Inhalation for 24 Months[*]: Adjusted Incidence of Squamous Cell Carcinoma of Nasal Cavity

Formaldehyde concentration (ppm)[†]	Number of tumors/animals at risk[‡]	
	Rat	Mouse
0	0/208 (0%)	0/72 (0%)
2	0/210 (0%)	0/64 (0%)
6	2/210 (1%)	0/73 (0%)
15	103/206 (50%)	2/60 (3.3%)

[*]6 h/day, 5 days/wk. The study was initiated with 240 Fischer-344 rats and 240 B6C3F1 mice, evenly divided by sex.

[†]Target concentrations. Actual average concentrations were 0, 2.0, 5.6, and 14.3 ppm.

[‡]Actual number of animals exposed to formaldehyde up to, and including, the interval when the first squamous cell carcinomas were found (months: 11–12 for rats; 23–24 for mice).

NASAL CAVITY SPECIFICITY

The exceptional water solubility and considerable reactivity of formaldehyde limits most exposure to the nasal cavity of obligatory nose-breathing species such as rats and mice. Autoradiography (see Swenberg et al., Chap. 12, this

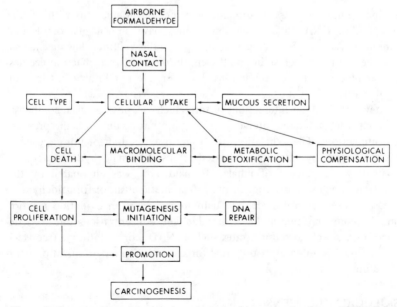

Figure 1 Conceptual model of various stages in the progression from the inhalation exposure of rats to formaldehyde to the development of squamous cell carcinoma in the nasal cavity.

volume) and classic disposition studies (see Heck et al., Chap. 4, this volume) using [^{14}C] formaldehyde demonstrate the localization of radioactivity in the nasal epithelium. Progressively smaller amounts reached the trachea, lung, and plasma during inhalation of [^{14}C] formaldehyde.

These results did not differentiate between cell types in the nasal epithelium nor did they take into account mucus or changes in the mucous lining overlying the epithelium. Since the nasal epithelium is comprised of a number of cell types, including ciliated and nonciliated cells (1), there may be differential sensitivity of cell types to formaldehyde. Conceivably this is an important effect, not yet addressed in the literature.

The respiratory tract mucus is comprised of large, extended glycoprotein molecules that have a high carbohydrate content. Cell origin, chronic lung disease, and various pharmacologic effects are known to influence the density of the anionic groups of these glycoproteins (2). The mucus may provide protection of respiratory epithelial cells at low concentrations of formaldehyde exposure, a protective mechanism lost at high formaldehyde concentrations. Some of the effects of formaldehyde may influence mucous secretion; and in the CIIT study, rats from all formaldehyde groups had goblet cell hyperplasia of the respiratory epithelium after 24 months (see Kerns et al., Chap. 11, this volume).

CELLULAR EFFECTS

Known biochemical effects of formaldehyde include interactions with cellular macromolecules. DNA alkaline elution studies have shown macromolecular cross-linking (3; see also Swenberg et al., Chap. 12, this volume). The mechanism involves the formation of stable methylene bridges between macromolecules. The consequences of these interactions, however, are not known but may include cell death.

Whatever the cellular interactions of formaldehyde, physiological compensatory responses or metabolic intervention may prevent or diminish the prospect of such interactions. The exhalation of $^{14}CO_2$ in expired air of rats exposed to [^{14}C] formaldehyde by inhalation indicated extensive formaldehyde metabolism. Approximately 40 percent of inhaled formaldehyde was eliminated by this route and presumably arose from formate. The metabolism of formaldehyde to formate, catalyzed by formaldehyde dehydrogenase, is an oxidation reaction that may become limiting at high formaldehyde concentrations due to a dependence on a supply of cosubstrates such as NAD+ or glutathione (see Heck et al., Chap. 4, this volume). A potential for saturation of a detoxification pathway therefore exists.

PHYSIOLOGICAL COMPENSATION

The most important observation concerning differences in the toxicologic response of rats and mice to formaldehyde involves the dose of formaldehyde reaching the nasal epithelium, the target tissue. Mice were more sensitive than rats to sensory irritation effects of formaldehyde so that concentrations of formaldehyde in air elicited a greater reflex slowing of respiratory rate in mice. Calculated on the basis of micrograms of formaldehyde reaching the nasal epithelium per minute per unit of surface area, rats received approximately twice the dose of mice at 15 ppm (see Barrow et al., Chap. 3, this volume). This difference in dose may account for the difference in apparent species sensitivity.

These observations suggest that the dose metameter used in risk analyses with formaldehyde should be expressed as an "effective dose" rather than as a concentration in air.

GENETIC EFFECTS

Formaldehyde was mutagenic in *Escherichia coli*, fungi, and L5178Y mouse lymphoma cells. However, no mutagenic response to formaldehyde was found in the CHO/HGPRT assay or in dominant lethal studies. Both positive and negative findings were obtained in Ames strains of *Salmonella*, depending on the conditions of the assay (4). In contrast to the mixed findings for bacterial and mammalian cell mutagenesis, however, formaldehyde was a pure initiator of carcinogenesis in the $10T\frac{1}{2}$ cell transformation assay (see Boreiko and Ragan, Chap. 7, this volume). A dose-related increase in transformed foci was found in the presence of a promoting agent, 12-0-tetradecanoyl phorbol-13-acetate (TPA). Formaldehyde, however, was not a complete carcinogen in these experiments since the presence of the promoter was required for transformation.

The above in vitro work indicated that DNA damage within cells of the nasal mucosa may occur in whole animals during formaldehyde exposure and that such damage may be fixed during cell replication. Opportunies for repair of damaged DNA exist, however, and DNA repair may modulate potential genetic injury at sites of damage. At low concentrations of formaldehyde, DNA repair may operate sufficiently to prevent fixation of chemical mutations. At higher concentrations of chemical exposure, repair may be saturated and unable to repair formaldehyde-induced DNA damage. Such mutational events may serve to initiate carcinogenesis, and in the presence of promotion, tumorigenesis may occur in the nasal epithelium. Conceivably one promotional process could be the dose-related increase in cell proliferation in the nasal epithelium of rats that accompanied exposure to high concentrations of formaldehyde (see Swenberg et al., Chap. 12, this volume).

A MULTISTAGE PROCESS

Carcinogenesis with formaldehyde is unlikely to be a one-hit process. Rather, as outlined above, it seems reasonable that many stages are involved in the process and some may operate to limit the effects of formaldehyde leading to carcinogenesis. Intuitively this must be true since formaldehyde is a normal endogenous metabolic product involved in one carbon metabolism (see Heck, Chap. 4, this volume).

RISK ASSESSMENT

Any risk analysis of formaldehyde-induced carcinogenicity must account for the various biological processes involved. It is important to seek a means of extrapolation from the high dose observed effects in rats to projected effects at low doses. Since a complete understanding of the mechanism of formaldehyde's carcinogenicity in rats is not known, no explicit statement concerning an appropriate model can be made. Many approaches for low dose extrapolation and risk assessment have been devised (5) and applied to the formaldehyde data presented above.

One approach is linear extrapolation. This approach considers that the probability of tumor formation is directly proportional to dose. Every dose, no matter how small, would be expected to carry some probability of tumor formation (6). In the case of formaldehyde, linear extrapolation into the low dose region is not appealing because formaldehyde is present in normal cells as an essential metabolic intermediate.

Other models, both statistical or stoichastic, have been suggested to obviate the inherent problems of linear extrapolation. Among these are the probit, logit, and Weibull distributions, as well as the gamma multihit and multistage models (7). None of these models, however, account for formaldehyde's effects explicitly at high and low concentrations and, therefore, at best could only approximate the expected response at low exposure concentrations.

It is instructive to consider the predicted concentrations of formaldehyde associated with the probability of tumor formation in rats using these models

Table 3 Virtually Safe Doses of Formaldehyde in Rats for Selected Excess Risks of Squamous Cell Carcinoma of the Nasal Cavity Predicted from Unadjusted Proportions (Table 1) Using the Probit, Gamma Multihit, Logit, Weibull, Multistage, and Linear Extrapolation Models

Model	Formaldehyde virtually safe dose (ppm)			
	1:100,000	1:1,000,000	1:10,000,000	1:100,000,000
Probit	2.56	2.08	1.73	1.46
Multihit	2.01	1.47	1.08	0.81
Logit	1.40	0.87	0.54	0.34
Weibull	1.27	0.76	0.46	0.28
Multistage	1.15	0.66	0.37	0.21
Linear	0.0066	0.00066	0.000066	0.0000066

Table 4 Virtually Safe Doses of Formaldehyde in Rats for Selected Excess Risks of Squamous Cell Carcinoma of the Nasal Cavity Predicted from Adjusted Proportions (Table 2) Using the Probit, Gamma Multihit, Logit, Weibull, Multistage, and Linear Extrapolation Models

	Formaldehyde virtually safe dose (ppm)			
Model	1:100,000	1:1,000,000	1:10,000,000	1:100,000,000
Probit	2.60	2.14	1.79	1.52
Multihit	2.05	1.52	1.14	0.86
Logit	1.41	0.89	0.56	0.35
Weibull	1.26	0.76	0.46	0.28
Multistage	1.26	0.65	0.37	0.21
Linear	0.0059	0.00059	0.000059	0.0000059

(Table 3). Five of the models essentially behave alike: the probit, logit, Weibull, gamma multihit, and multistage. The linear extrapolation model stands by itself. The same general conclusions are reached whether the analyses are conducted on the absolute number of tumors observed in all of the animals started on study and examined (Table 3) or on the adjusted data wherein the incidence of tumors is expressed per number of animals at risk (Table 4). When considering low dose extrapolation it is important to keep in perspective that outdoor formaldehyde concentrations may be in the range of 0.05 ppm (8). However, no absolute increases in nasal cancer were found in rats exposed to 2 ppm formaldehyde for 24 months or to 6 ppm in mice for the same period.

Low dose extrapolation of the CIIT rat and mouse data for the purpose of human risk assessment must await definitive knowledge of the mechanism of formaldehyde's effects. In addition, specific information on endogenous concentrations of formaldehyde in human tissues, cytological effects, occupational health effects, and mortality experience in humans must be gained to fully understand formaldehyde's potential for carcinogenesis in humans.

The significance of the carcinogenic effects in rats, however, dictate that formaldehyde be kept at the lowest practical concentration in the workplace, in homes, and in ambient air. The concentrations of formaldehyde suggested by linear extrapolation of the rodent data are really not practical. More meaningful risk assessments and extrapolations must await further information on the mechanism of formaldehyde's carcinogenic effect in these animals.

REFERENCES

1 Gross, EA, JA Swenberg, S Fields, and JA Popp. Comparative morphometry of the nasal cavity of rats and mice. *J. Anat.* 135:83–88 (1982).

2 Clark, JN, WE Dalbey, and KB Stephenson. Effect of sulfur dioxide on the morphology and mucin biosynthesis by the rat trachea. *J. Environ. Pathol. Toxicol.* 4:197–207 (1980).

3 Swenberg, JA. Utilization of the alkaline elution assay as a short-term test for chemical carcinogens. In *Short-Term Tests for Chemical Carcinogens*, edited by HF Stich and RHC San, pp. 48–58. New York: Springer-Verlag (1980).

4 Boreiko, CJ, DB Couch, and JA Swenberg. Mutagenic and carcinogenic effects of formaldehyde. In *Genotoxic Effects of Airborne Agents*, edited by RR Tice, DL Costa, and KM Staich, pp. 353–367. New York: Plenum (1982).

5 Krewski, D, and B Brown. Carcinogenic risk assessment: A guide to the literature. *Biometrics* 37:353–366 (1981).

6 Gaylor, DW, and RL Kodell. Linear interpolation algorithm for low dose risk assessment of toxic substances. *J. Environ. Pathol. Toxicol.* 4:305–312 (1980).

7 Krewski, D, and J Van Ryzin. Dose response models for quantal response toxicity data. In *Statistics and Related Topics*, edited by M Csorgo, D Dawson, JNK Rao, and E Saleh, pp. 201–231. Amsterdam: North-Holland (1981).

8 Stupfel, M. Recent advances in investigations of toxicity of automotive exhaust. *Environ. Health Perspect.* 17:253–285 (1976).

Chapter 27

Concluding Remarks and Recommendations for Future Research

Leon Golberg

The foregoing chapters have presented a great deal of information—both old and new. It all needs to be assembled, digested, assimilated, and weighed carefully before decisions can be reached on the basis of these data. While the research has answered many questions, it also has opened up fresh vistas—intriguing issues of great practical importance that some of us might not have thought about before. Here indeed is a case in which "what's past is prologue."

In this chapter I will review briefly the issues that may be regarded as settled, point out the areas of residual uncertainty, and outline the recommendations for future research that may be stimulated as a consequence of these studies.

Out first concern is, of course, with the test substance: has the work with formaldehyde, whether in gaseous form or applied as a solution, used appropriately pure material? I think we can feel comfortable on this score. The next question is: were the effects observed caused by the test substance? A cautious "yes" is probably in order, provided that we leave open the distinction between direct effects of formaldehyde itself, or a reactive moiety derived from it, and effects mediated by one or more interaction products of formaldehyde with components of nasal mucus or of subjacent or other tissues. Whatever the active entity or entities may be, primary tissue irritation at the point of impact and beyond is the initial effect, followed by inflammation, rhinitis, necrosis of nasal columnar or cuboidal respiratory epithelium, and sloughing of the necrotic material. Apparently the squamous respiratory epithelium and the olfactory epithelium are less readily damaged. In the rat, and probably other species, reparative restoration of the injured mucosa is remarkably swift and complete.

Adaptive development of tolerance to the irritant and cytotoxic properties of formaldehyde leads to some striking differences between the responses of naive and previously exposed animals. Nevertheless, in both categories of animals, continued exposure to high levels of formaldehyde sets in motion a sequence of pathogenesis that then proceeds via epithelial dysplasia to squamous metaplasia. Repair and repopulation of the nasal mucosa seems thus directed to lining the surface with a more resistant population of squamous cells actively producing keratin. What triggers the progression to neoplasia—squamous or adenomatous—remains uncertain. Especially important is the evidence that nasal epithelial changes up to and including squamous metaplasia are fairly readily reversible on cessation of exposure, except where the high level and long duration of exposure has taken the pathological process too far.

Even though the results of current studies are still incomplete, some tentative general conclusions may be perceived. Neoplasia in the rat and mouse shows a dose-response relationship. What appears to be a greater susceptibility on the part of the rat may be, at least partially, a consequence of the fact that the mouse is subjected to about one half the target dose of formaldehyde as the rat when the two species are exposed to the same levels in the air. Acceptance of this view in turn greatly diminishes the possible responsibility of sialodacryoadenitis for eliciting neoplasia in the rat.

We have the interesting evidence that while formaldehyde is an initiator of neoplasia, promotion by some other agent or factor can make a substantial difference to the capacity to induce neoplastic transformation in vitro. In the course of inhalation of levels as high as 15 ppm formaldehyde, promotion very likely takes the form of intense cell replication—at 70 times the normal rate—set in motion through the havoc wreaked by formaldehyde in susceptible epithelia. If cell replication is indeed the promoting influence, then the concentration of formaldehyde in air is the important, perhaps the primary determining factor in the steep dose-response curve, rather than the cumulative tissue dose. First it is concentration that determines cytotoxicity. There is little or no prospect of the aldehyde lingering in the tissues in free form. Relatively minor degrees of tissue damage are probably repaired during the daily periods of nonexposure or over the weekends. Overall, therefore, the picture is one of dynamic reversibility of or tolerance development to irritant effects at low levels of exposure; less rapid but probably still completely effective regression of more advanced pathological changes; and, at high levels for long periods, even lesser degrees of reversibility or none at all. This is a familiar picture that has been seen, for example, in the bladder mucosal hyperplasia of rats exposed to carcinogens like FANFT.

We must, however, ask: is there a residual tissue "memory" of exposure to formaldehyde, despite the apparently complete restoration to normal? In the studies of genetic toxicity we have evidence that formaldehyde can induce both point mutations and chromosomal aberrations in appropriate systems. It seems evident that formaldehyde or some product derived from it reaches the germinal cells of animals. When exposure takes place to water-soluble, labile sources of formaldehyde such as hexametylenetetramine or dinitrosopentamethylenetetramine, nothing much happens. But stable lipophilic sources such as hexamethylphosphoramide slip into the cell like the wooden horse taken into Troy and, once inside, slowly release formaldehyde with devastating effects. The classic experiments with *Drosophila*, yielding positive results when the formaldehyde was given in food but not in aqueous solution or gaseous form, suggest caution in constructing an Amesian equation: positive mutagenicity, ergo positive carcinogenicity. I feel that the potential for carcinogenicity is probably expressed only under appropriate circumstances. Also, some weight should be given to the fact that formaldehyde has not proved to be teratogenic in well-conducted experiments in mice. Studies in rats and dogs were not thorough enough to permit a conclusion to be drawn from them.

Turning to human exposure, clearly these studies are in their infancy. In fact, the suggestion that infants are more sensitive to formaldehyde than adults demands further exploration. There is a need to look at the incidence of spontaneous abortions. I am wondering, too, whether the nasal cavities could be investigated in people exposed occupationally to formaldehyde—pathologists, histotechnicians, embalmers. Biopsy of the nasal mucosa has proved a valuable diagnostic tool in nickel workers. Study of the cytology of nasal secretions in formaldehyde-exposed workers may also be useful. Maibach set forth several urgent needs in the area of cutaneous effects (see Chap. 15, this volume). What comes through loud and clear is the need for better indexes of exposure and more objective means of assessing effects in humans.

Finally, both in humans and in animals, we have an outstanding example of the need to study combined effects, especially with other respiratory irritants and other potential interactions of formaldehyde in everyday life.

Index